THE AWFUL TRUTH

by Patrick J. Conway

To Jax

Enjoy

Pat Conway

Published by Pragmatic Press
P. O. Box 260199, Bellerose, New York 11426-0199

Publisher's Cataloging-in-Publication Data
Conway, Patrick J.
The awful truth: for those who dare to
know / Patrick J. Conway. —Bellerose, NY:
Pragmatic Press, 2000.
p. cm.
Includes bibliographical references and index.
ISBN 1-880959-39-9
1. Apocalyptic literature—Comparative studies. 2. End of the
world—Comparative studies. 3. Millennialism. I Title.
BL501 .C66 2000
291.2/3—dc21   CIP

Printed in the United States of America

02   01   00   99 •   5   4   3   2   1

The author makes grateful acknowledgment to the following:

The Associated Press, for permission to quote from "Noble Lie" by religion writer George W. Cornel; dated 19 July, 1991. ©1991 Associated Press

U.S. News & World Report, for permission to quote from "A No-Fault Holocaust" by John Leo; dated 21 July, 1997. ©1997 U.S. News & World Report.

Harper Collins Publishers, for permission to quote from *Secrets of the Great Pyramid* by Peter Tompkins. ©1971

International Publishers, for permission to quote from *The Origins of Christianity* by Archibald Robertson. ©1954

Adler & Adler, for permission to quote from *Evolution: A Theory in Crisis* by Michael Denton. ©1985

Kalev Pehme, for permission to quote from his lecture "The Myth of Race," delivered at the Henry George School on 4 February, 1994.

Tina Collin for cover design

Mary Horenkamp for layout

Claudio Octaviano for illustrations

And to the late John Grella and his family for seeing the value of my thoughts and buying the first copy of this book seven years before it was ever completed.

This book is dedicated to my wife, Josephine, and my three children: Patrick, Kevin and Megan. Your willingness to endure years of slowly depleting financial resources to produce this work is the true meaning of the word *love*.

Josephine—"Maybe you should write a book!" I am certain that in the fifteen years since you made that simple-sounding suggestion you have had many regrets. It is a testimony to your love and dedication to our marriage, family and the theory that you never voiced them to me. Your input and criticisms were, in a single word, invaluable.

Patrick—An editor's job is to criticize and discipline his client. It requires the skills of a teacher, disciplinarian and diplomat. To assume the position of editor to your father is to risk all. Spending three years in that role and successfully completing the task constitute a monumental labor of love. Without your selfless sacrifice, this book would not exist.

Finally, I dedicate this book to myself. Let it never be forgotten that it is the result of twenty-five years of exhaustive research in the face of stiff opposition from both narrow-minded religious groups and virulent atheists in academia. In addition, there is a great personal price to be paid for taking a stand in opposition to the ideological beliefs of one's family and friends. Had I known the full extent of those costs when I started, this book would never have been written.

Table of Contents

	Introduction	*vii*
I.	Apes or Aliens?	1
II.	The Fall of the Gods	29
III.	Humans Inherit The Earth	105
IV.	We Can't All Be Right	191
V.	Helping The Believers Be Honest	203
VI.	Paradigm Lock	217
VII.	No Place To Hide	255

INTRODUCTION

The phrase *an awful truth* describes newly discovered facts that expose an error in our current thinking. We use the word *awful*, even in cases where the new reality produced by such a revelation is beneficial, because innovations that challenge our core beliefs frighten us. This apprehension stems not from the truth itself; it is our fear of change that makes even a benevolent new concept seem awful.

For instance, prior to the Renaissance, scientists mistakenly believed that the earth was located at the center of the universe. The Church[1] hierarchy adopted this geocentric theory, and it became a tenet of early Christianity. When the Italian astronomer Galileo produced evidence for the heliocentric, or sun-centered, solar system, Church scientists were confronted with the awful truth that the earth was a planet orbiting the sun. Unwilling to relinquish their long-held belief, these scientists rejected the evidence and continued to support the geocentric system. They then conspired with the Church fathers, who feared that this contradiction of dogma would weaken Christian faith, and ultimately had Galileo censured by the Inquisition. Nevertheless, before long even the Church scientists were forced to accept the reality of the sun-centered solar system. When they did, the "awful truth" they had so feared produced a truly benevolent reality that has allowed man to calibrate an accurate calendar, optimize his use of the agricultural seasons and explore the globe. In addition, contrary to the irrational fears of the Church fathers, accurate knowledge of this delicately balanced system has only increased mankind's awe for its Creator. So you can see that an awful truth, or the discovery of disturbing new facts that force us to correct erroneous thinking, will always prove to be beneficial.

That is the type of benevolent truth revealed in this book. Yet, because it contradicts the orthodox beliefs of science and religion, it will, no doubt, stir the irrational fears of experts in both fields. Nevertheless, I am convinced that what you are about to read will solve the great mystery of our existence and, like the discovery of the heliocentric solar system, advance the knowledge and well-being of all mankind.

1. Apes or Aliens?

Most ancient civilizations, both biblical and pagan, have historical records stating that supernatural beings transported our ancestors to the earth from distant galaxies. These texts and traditions disagree about the precise location of our origins, but they universally describe man as an alien. For instance, the Inca tradition states that their ancestors came from the constellation of Pleiades. Hopi Indians in the American southwest say that their ancestors came from endless space and that they had visited many worlds before settling on Earth. The Mayan gods reportedly came from the stars, and the *Popol Vuh* creation legend of the Quiche Maya states that four hundred heavenly beings left earth and returned to Pleiades. In the case of ancient Israel, biblical history records that man was created in a place called the Garden of Eden and later placed on the earth. Unsure of the exact location of Eden, early Church leaders allowed that the Garden described in Genesis was probably located in some yet unexplored, remote corner of the planet. This position avoided controversy, but it totally ignored the extraterrestrial implications of the Bible's account of our creation. Scientists compounded the problem when they adopted the inherently flawed theory of evolution, discounting as myth the biblical and pagan accounts of our extraterrestrial origins.

Though science and religion disagree over the genesis of man, in recent years some denominations of Judeo-Christianity[1] have given tacit approval to the scientific position that those origins are rooted in the earth. Unfortunately for both disciplines, the ecological evidence contradicts that belief. All earthly creatures live in complete harmony with nature, with the singular exception of humans. Our lifestyle is totally alien to Earth's environment. Having no natural fur to protect us from the elements, we must take our clothing from other creatures or manufacture it synthetically. Uncomfortable in any climate, we build houses, then heat them in winter and cool them in summer, consuming large quantities of the earth's resources. Subhuman creatures have natural predators designed to help keep their populations in check; man has none. Beyond that, man is endowed with a higher intelligence, and we alone have dominion over the planet's food supplies. Humans also possess the ability to utilize medical skills to extend our life spans beyond the bounds that nature would normally permit. Unlike the animals, which become fertile only once a year, humans can conceive monthly; consequently, we are overpopulating the planet, depleting the earth's resources and polluting the environment. The earth's ecology is delicately balanced on a system of death. All creatures, with the exception of humans, live in harmony with that natural law, which recycles the planet's resources through nature's crucible of the survival of the fittest. This system is designed to prevent creatures that are indigenous to the planet from damaging the environment through overpopulation and pollution. If an animal produces a weak offspring, within a short period of time a predator will kill it, and the herd goes on, unmoved by this perfectly natural event. On the other hand, humans are dedicated to the survival of the weakest. We will take the most premature baby, rush it to an incubator and try to nurse it into life, and we will go to equal lengths for an eighty-year-old cancer patient. If the herculean efforts of the medical community fail and the patient dies, we go into a period of mourning because we find death abhorrent. Unlike the animals, we humans possess an innate sense of immortality. Our relentless efforts to sustain life and great fear of death stand in direct opposition to the basic principles of the earth's ecology. Thus, the very essence of human nature is antithetical to the Darwinian concept that species evolve to better suit their environments. A primate who lived in perfect harmony with the ecology of death would not evolve to find it abhorrent. The preponderance of ecological and sociological evidence reveals that man is an extraterrestrial being whose origins are not located on the earth. As we proceed, you will see that the awful truth inherent in that reality has ramifications that are far more beneficial to mankind than anything yet suggested in the erroneous paradigms of science and religion.

SCIENCE VS RELIGION
While They Can't Both Be Right, They Can Both Be Wrong!

In the early modern world, as man began to try to understand his environment, an ideological dispute erupted between science and religion. That conflict is at the heart of our inability to solve the mysteries of both our origins and the great wonders of the highly sophisticated civilizations of the ancient world. If we are to work together to answer the salient questions surrounding those mysteries, we must revisit the cause of that conflict.

At the beginning of the dark ages,[2] all men believed in God, and science was part of theology. Early theological science was, for the most part, sponsored by the Christian Church, and it was defined as the use of logic, tempered by the moral constraints of Scripture, to try to understand God and His Creation. Though the essence of theology was noble in theory, the same cannot be said of its practitioners. In the 16th century, some theological scientists, such as Galileo, began to make discoveries that contradicted Church dogma. Fearing that these new revelations would weaken Christian faith, the hierarchy simply rejected them. Church scientists were forced to ignore contrary discoveries and faithfully follow Christian teaching. In this manner, the Church fathers exempted themselves from the constraints of logic, and Christianity became a cult of moralistic believers.

No longer governed by the logical aspect of theology, Church leaders could now require the faithful to believe doctrines that had no basis in reality. Objective scientists like Galileo, who remained loyal to the true principles of theological research, were persecuted. Though these ostracized theologians were forced to work apart from the Church, they continued to believe in God and studied the ancient texts, both biblical and pagan, to increase their knowledge of mankind's place in the universe. These texts reveal that all ancient science was based on the extraterrestrial origins of man. Though they differ in detail, they universally state that the supernatural beings who ruled the earth in the ancient world had either created mankind here on earth or transported us to the planet at the dawn of human history. In either case, man is described as a nonindigenous creature. Notwithstanding, as the intellectual acrimony between the ostracized scientists and the Church hierarchy intensified during the Renaissance, these dissidents began to seek a more earthly explanation for man's origins as a means of breaking the clergy's dictatorial control over academic thought. In the 19th century, these separated scientists adopted Darwin's theory of evolution as a means to that end, and this ideological split became a chasm.

Darwin's Theory Turns History Into Myth

Slowly, evolution became the cohesive center around which the new secular science coalesced. Unlike earlier dissidents, Darwinian scientists attacked the credibility of the Bible itself by using evolution to discredit its core tenet—the concept that mankind and the universe were created by an Almighty God. It cannot be claimed that the atheists were unaware of the flaws in Darwin's theory; in fact, it was roundly criticized by many of his peers. Nevertheless, the more radical among these dissidents were determined to use it to expose the inconsistencies in biblical teaching and thereby discredit the Church. Their great fear of papal tyranny and fanatic devotion to Darwin drove them to extremes. They rejected out-of-hand thousands of years of biblical and pagan history in which our ancestors categorically stated that man was a created alien who was placed on the earth by the supernatural forces that ruled the planet in the ancient world. This great wealth of knowledge was simply cast aside as myth, and the advocacy of atheistic evolution became a badge of intellectual acumen among academics. Scientists joined with other atheistic groups in an effort to debunk the Bible and were successful in portraying their theological colleagues as neanderthalic simpletons. In this manner they separated themselves from the constraints of both history and biblical morals and, in effect, became an atheistic cult. As a result, all hope of reunification with the Church scientists ended.

Over time, these atheists became the dominant force in academia, forming a very authoritarian caste which assumed dictatorial powers not unlike those of the old Church hierarchy that they had so despised. With the roles reversed, scholars in any discipline who even questioned atheistic dogma or dared to support the ancient historical texts in any way were ostracized, and their research was summarily rejected. As the Church had exempted itself from the constraints of logic by declaring that the human intellect was incapable of understanding God and His Creation, secular science had now made the equally tragic mistake of declaring itself exempt from the constraints of both history and "biblical morals." As a result, the guiding principles essential to an objective search for truth—logic and morality—were permanently divided between two cults of believers. Dividing these essential parts of the whole between science and religion made it literally impossible to understand either God or man. Science declared the historical accounts of man's intercourse with the gods to be the illogical or "mythical" delusions of Neanderthals, and religion declared all atheistic science, regardless of how logical, to be immoral or "satanic." Later, we will see how this ideological rift has created our current political divisions of Left and Right. Without a clear understanding of the causes and consequences of this Church-science jihad, we will never be able to find a common ground from which to heal our ideological wounds and become a united society.

Uncomfortable with the thought of being described as immoral, secular scientists invented the word *amoral* to describe their atheistic research. The dictionary defines this word as "not subject to or concerned with moral or ethical distinctions." This is a totally fallacious term. If the clergy were to follow this example and make up a word like *alogical* to excuse their unwillingness to subject religious beliefs to the constraints of logic, these same scientists would scoff at such a concept. All things are perceived in both a moral and logical context. With each of these philosophical cults ignoring one of the prerequisite elements of pragmatic research, science and religion have concocted half-baked paradigms on the existence of man that are both illogical and immoral. It is illogical to believe that the gods of the ancient world are active in the affairs of modern man, and it is immoral to subject children to the psychologically destructive belief that they are descendants of the primates. Human beings are godlike creatures whose rationalism is tempered by ethics; therefore, a combination of both principles is needed to determine the origins of such a noble creature.

As the followers of these cults became more radical, internal disputes split them even further, and they spawned literally thousands of new cults, leaving mankind divided and confused. In the madness of it all, we seem to have lost sight of the fact that the original argument over the origins of man was never settled. Scientists continue to claim that man is a descendant of the primates, while the Church holds that we were created by an Almighty God. In recent years, these antagonists have been trying to pretend that somehow they are both right; that although man is a serendipitous byproduct of evolution, the process may have been planned by a Creator. This, too, is an outright denial of reality. These totally polarized models of the origins of man cannot both be right. We must determine which paradigm is true—the evolving primate of science or the created alien of the ancient texts.

The leaders of these cults use self-serving rhetoric to convince us that their ideology is, on the one hand, more "holy" and, on the other, more "objective." In addition, they bestow great honors and titles on each other like "Doctor of Theology" or "Doctor of Science" to enhance their credibility. This propaganda only muddies the water; I ask the reader to ignore the hype and identify these ideologues in accordance with their respective amoral or alogical beliefs.[3] We will see that when it comes to either holiness or objectivity, both sides receive a failing grade. In reality, they are all part of the intelligentsia, and their ideological divorce would best be described as a hissy fit among the elite of the overarching cult of academia.

Belief vs Truth

When an erroneous belief or hypothesis goes unchallenged for a long period of time, as did the geocentric universe, its advocates begin to think of it as a truth. A major cause of this confusion is the misapplication of the word *belief*. The dictionary definition of this word reads as follows: "acceptance of the truth or actuality of anything without certain proof." This is a very dangerous concept because it encourages us to accept the illusion that a belief can be true even though it is totally devoid of fact. Once we accept this patently false premise, we can be easily convinced to follow doctrines that contradict reality and defy our innate sense of logic. Clergymen may require us to pray to a piece of bread, fight and die for an ancient homeland besieged by a jihad, become a human bomb or handle poisonous snakes in order to please God. On the other hand, atheistic scientists may require us to believe that all life evolved from one primeval cell, that humans are slowly evolving primates or that they can determine the age of a rock that is "billions" of years old. They may even ask us to believe that the solar system is the result of a spontaneous mixing of gases in the great void of space and that there is no intelligence behind this magnificent and intricately balanced network of planets. What's worse, these professionals often take offense when asked to justify their unsubstantiated beliefs. Both sides seem to presume that their academic positions place them above the level of questions generated in the logic of the common man. The time has come to confront these professionals with the madness of their current beliefs and force them to face the awful truth—that neither side has solved the great mystery of man's existence.

Truth: What Is It?

Unlike a belief or hypothesis, of which there can be countless versions, there is only one truth to any given question. In fact, there are very few things that we all accept to be true, while we have literally thousands of beliefs on many subjects. A truth is based in fact. The next question is, Whose facts? But even this good question is flawed. Truth is not personal, it's universal, and the facts that support a truth will be both understood and accepted by everyone. Take death, for instance. No one asks whose facts you are using when it comes to the subject of death. When asked if you are going to die, you don't answer by saying, "I believe in death" because death is a universal truth. Another example is the solar system. No one says that they believe in the heliocentric solar system; we just acquiesce to it as a truth. Prior to the discovery of the Copernican model, there were many beliefs about the earth and its relationship to the sun and the other planets. For example, some believed that the earth traveled through space on the back of a turtle, some that it was flat, and most early modern scientists believed that it was located at the center of the universe. When the solar system's true configuration was finally discovered by Copernicus, all of these mistaken beliefs eventually fell by the wayside. The same will be true for our current beliefs about God and the origins of man; when the truth of these matters is finally discovered, all of our erroneous scientific and religious theories will be swept away. The very fact that we have literally thousands of conflicting scientific and religious beliefs about those mysteries is evidence that we have not, as yet, discovered the truth. To avoid the dangerous misconceptions that result when a long-held belief evolves into an accepted scientific or religious doctrine, the words *belief* and *hypothesis* should be totally separated from any connotation implied by the word *truth*. Truth is not the private domain of either science or religion. The fact that science has concocted a paradigm does not automatically make it factual, nor is a time-honored religious belief necessarily endorsed by God. The reality that neither science nor religion has produced a universally accepted truth on the origins of man declares the weakness of their current positions.

The State of Denial

In addition to their propensity to blur the line between hypothesis and truth, atheistic scientists will simply deny reality when a current belief is proven to be erroneous. This problem has its roots in professional pride, and contemporary scientists are no better than the early Church scientists when it comes to dealing with an awful truth. In his eye-opening book *Evolution: A Theory In Crisis*, Dr. Michael Denton exposes the state of denial surrounding the growing crisis in evolutionary science. Recent discoveries in biology and paleontology have made it impossible to continue to excuse the conspicuous flaws that have plagued the theory of evolution since its inception. Nevertheless, most scientists continue to deny the evidence in order to protect their enormous investment in Darwin.

Dr. Denton's work is exemplary because he does not espouse any alternative to evolution. He merely uses 20th century biology and paleontology to debunk Darwin's basic premise—that all life evolved from lower forms of existence. Dr. Denton is neither a creationist nor a Bible-thumping Christian, but rather an objective scientist who used only empirical evidence as a foundation for his work. Consequently, his fellow scientists were at a loss as to how to deal with his book. Unlike the Christian fundamentalists, who provide scientists with a long list of indefensible religious doctrines to attack, Dr. Denton did not have an agenda. This left scientists in the uncomfortable position of having to defend Darwin on the merits. To avoid that dilemma, they tried to discredit Dr. Denton by subjecting him to personal attacks and charges of quackery, but ten years after publication, his facts stand unchallenged.

The flaws inherent in the Darwinian philosophy were known from the very beginning, but modern technology has amplified those discrepancies to the point where even a layman like myself can be made aghast at the scope of this incredible "scientific" myth. Speaking of these flaws, Dr. Denton writes:

> Neither of the two fundamental axioms of Darwin's macroevolutionary theory—the concept of the continuity of nature, that is the idea of a functional continuum of all life forms linking all species together and ultimately leading back to a primeval cell, and the belief that all of the adaptive design of life has resulted from a blind random process—have been validated by one single empirical discovery or scientific advance since 1859. Despite more than a century of intensive effort on the part of evolutionary biologists, the major objections raised by Darwin's critics such as Agassiz, Picket, Bronn and Richard Owen have not been met. The mind must still fill up the "large blanks" that Darwin acknowledged in his letter to Asa Gray.

> One hundred and twenty years ago it was possible for a sceptic to be forgiving, to give Darwin the benefit of the doubt and allow that perhaps future discoveries would eventually fill in the blanks that were so apparent in 1859. Such a position is far less tenable today. (Denton, p.345)

What are the objections to Darwin's theory of evolution? Dr. Denton's book reveals so many flaws that it is difficult to know where to start, but the most obvious and easily understood discrepancy is the total lack of evidence in the fossil record. Darwin's theory of the gradual evolution of species, say, from fish to amphibian, would require a long period of transition. As Darwin himself stated, for the theory to be valid there should be a wealth of transitional fossils revealing that change, and this should be true for all other species.

No transitional fossils had been found at the time Darwin published *Origin of Species*, and this led to much criticism of his work by fellow scientists. Darwin attributed this lack of evidence to the limited ability of early science to explore the planet, and he assured his critics that these blanks would be filled by future scientific discoveries. Had he been right, by now we should be literally drowning in transitional fossils, but the dirty little secret of science is that there are no transitional fossils. None. Nada. Zero. Zip. Zilch. Of all the multiplied millions of fossils discovered since the time of Darwin, not one can be labeled as a transitional fossil.

In an effort to save their cherished theory of evolution from total censure, scientists have recently developed a variation called *punctuated equilibrium*.[4] This theory was put forth by American paleontologists Niles Eldredge and Stephen Jay Gould. It postulates that evolution occurs in short bursts of rapid change and is followed by long periods of stability. They further speculate that because these transitional spurts produced very small populations in isolated regions of the planet, those species died off without leaving a fossil record. While this does conceivably provide an explanation for the lack of transitional fossils, it provides no positive evidence for evolution. On the other hand, the fact that scientists have to concoct this contrived explanation only highlights the problem, allowing opponents to legitimately claim that the absence of fossils is more easily explained by admitting that evolution is a myth. On the more scientific side, Dr. Denton writes that while punctuated equilibrium might be said to explain the lack of transitional fossils between species, it cannot be extended to solve the larger, systematic gaps:

> The gaps which separate species: dog/fox, rat/mouse etc. are utterly trivial compared with, say, that between a primitive terrestrial mammal and a whale or a primitive terrestrial reptile and an Ichthyosaur; and even these relatively major discontinuities are trivial alongside those which divide major phyla such as molluscs and arthropods. Such major discontinuities simply could not, unless we are to believe in miracles, have been crossed in geologically short periods of time through one or two transitional species occupying restricted geographical areas. Surely, such transitions must have involved long lineages including many collateral lines of hundreds or probably thousands of transitional species.
>
> To suggest that the hundreds, thousands or possibly even millions of transitional species which must have existed in the interval between vastly dissimilar types were all unsuccessful species occupying isolated areas and having very small population numbers is verging on the incredible! (Denton, p.193)

These desperate attempts to save evolution are similar to the absurd lengths that early scientists went to when trying to save Ptolemy's geocentric universe. When observations of the moving planets didn't fit the standard geocentric model, scientists contrived a complex system of epicycles[5] to try to explain the contradictory evidence. They continued this futile effort for more than one hundred years, until the undeniable evidence supporting the heliocentric solar system forced them to face reality. Modern scientists are making the same mistake with Darwin's model of evolution. By concocting theories like punctuated equilibrium, they only postpone the inevitable.

The effort by Eldredge and Gould to save Darwin's dying theory has only served to awaken both the public and the scientific community to the lack of supporting evidence in the fossil record. Speaking of the magnitude of the problem, Dr. Denton says:

> The advent of the theory of punctuated equilibrium and the associated publicity it has generated have meant that for the first time biologists with little knowledge of paleontology have become aware of the absence of transitional forms. After this revelation of what Gould has called "the trade secret of paleontology" it seems unlikely that we will see any return in the future to the old comfortable notion that fossils provide evidence of gradual evolutionary change. (Denton, p.194)

This is an absolutely astonishing statement. Translation: Darwin's model for the origins of life is finished.

I have always been aware of evolution's incompatibility with logic, but I was shocked by the glaring biological flaws exposed by Dr. Denton. They leave one bewildered as to how scientists could have allowed this sham to become the dominant tenet of Western science. It appears that biologists, like the paleontologists, ignored the scientific flaws in evolution hoping that future discoveries would solve the molecular problems. That hope was dashed by the technological advances of 20th century. This technology revealed the complex and systematic nature of cellular structure, making it all but impossible to pretend that the vast and complex array of living organisms on earth are the result of evolution. Undaunted by the evidence, scientists continue to believe that the genetic structures of living organisms are the result of a serendipitous accident of nature. In effect, they are saying that their minds are made up and that they don't want to be confused by the facts. Dr. Denton writes:

> To the sceptic, the proposition that the genetic programmes of higher organisms, consisting of something close to a thousand million bits of information, equivalent to the sequence of letters in a

small library of one thousand volumes, containing in encoded form countless thousands of intricate algorithms controlling, specifying and ordering the growth and development of billions of cells into the form of a complex organism, were composed by a purely random process is simply an affront to reason. But to the Darwinist the idea is accepted without a ripple of doubt—the paradigm takes precedence! (Denton, p.351)

In his chapter "The Priority of the Paradigm," Dr. Denton makes the case that scientists will defend a flawed theory, regardless of the volume of contrary evidence, until forced to change by the emergence of a new paradigm. He cites the fact that scientists held fast to the flawed medieval theories on phlogiston combustion[6] and the geocentric universe, refusing to relinquish them in the face of overwhelming contradictory facts. In other words, like the leaders of any cult, scientists will defend their erroneous beliefs until the rest of the world forces them to change by embracing an undeniable new reality.

Astonished by the fact that Darwinian evolution became the dominant tenet of scientific thought, driving all other theories from contention, Dr. Denton says:

> One might have expected that a theory of such cardinal importance, a theory that literally changed the world, would have been something more than metaphysics, something more than myth.

> Ultimately the Darwinian theory of evolution is no more nor less than the great cosmogenic myth of the twentieth century. (Denton, p.358)

Who was Darwin that we felt compelled to throw away thousands of years of human literature, history and tradition from around the world and evict the Bible from our schools to accommodate his intrinsically flawed theory? No hypothesis in history has had such a profoundly negative effect on the modern world as has atheistic evolution. It not only caused a totally unwarranted intellectual revolution, but later we will see that it has turned our schoolchildren into ideological pawns in the extremely destructive conflict between science and religion.

I think that the following quote from *Descent of Man* will give the reader some insight into Darwin and the mind-set of science in the 1800s.

> I now admit, after reading the essay by Nageli on plants, and the remarks by various authors with respect to animals, more especially those recently made by Professor Broca, that in the earlier editions of my Origin of Species I probably attributed too much to the action of natural selection or the survival of the fittest. I have altered the fifth edition of the Origin so as to confine my remarks to adaptive changes of the structure. I had not formerly sufficiently considered the existence of many structures which appear to be, as far as we can judge, neither beneficial nor injurious; and this I believe to be one of the greatest oversights as yet detected in my work. I may be permitted to say as some excuse, that I had two distinct objects in view, firstly to shew that species had not been separately created, and secondly, that natural selection had been the chief agent of change, though largely aided by the inherited effects of habit, and slightly by the direct action of the surrounding conditions. Nevertheless, I was not able to annul the influence of my former belief, then widely prevalent, that each species had been purposely created; and this led to my tacitly assuming that every detail of structure, excepting rudiments, was of some special, though unrecognised, service. Any one with this assumption in his mind would naturally extend the action of natural selection, either during past or present times, too far. Some of those who admit the principle of evolution, but reject natural selection, seem to forget, when criticising my book, that I had the above two objects in view; hence if

> I have erred in giving to natural selection great power,
> which I am far from admitting, or in having
> exaggerated its power, which is in itself probable, I
> have at least, as I hope, done good service in aiding to
> overthrow the dogma of separate creations.
> (Darwin, p.152-3; Ch. 4, part 1)

This quote is very revealing on several counts. First, we see the great controversy surrounding Darwin; his theories were roundly criticized by his contemporaries in secular science. Second, Darwin reveals a sizable ego by only half-admitting to his errors with regard to natural selection and then blaming others for taking his work too literally. Later on we will see that we can add Nazi scientists to the list of people who made that tragic mistake. Finally, Darwin suggests that his fellow scientists should overlook the errors in his theories because they have helped to destroy the concept of separate creations. In other words, it was quite all right to fudge on the facts as long as scientists prevailed in their ideological struggle with the clergy over the origins of life. Disgraceful! At one time it may have been tolerable to allow for some flaws in the theory of evolution; after one hundred thirty-five years, it is inexcusable. We must decide: Is the scientific paradigm of evolution a truth or a hoax?

Dr. Denton's chapter on the priority of the paradigm should be must reading for every high school student. It helps us to see how experts in any field can cause an erroneous concept to dominate the philosophy of an entire civilization simply by refusing to address evidence hostile to their beliefs. This knowledge will be invaluable to these young students when confronted with similar problems in other professions.

Religion is another place where the drive to protect the paradigm in the face of blatant contrary evidence is legendary. I recently saw an obvious case of this on national television when the Reverend Billy Graham addressed a memorial service for the victims of the terrorist bombing that destroyed the Alfred P. Murrah Federal Building in Oklahoma City. He stated that during his many years of ministry he had been asked to visit several disasters, both natural and man-made, and always the question of *why* arises. Why does God allow seemingly good and innocent people, even little children, to become the victims of senseless evil? He admitted that on several occasions circumstances in his own life had forced him to ask the same question and that he did not know the answer. Nevertheless, he said that his faith led him to believe that God was a merciful and loving God who cared for him on a personal level and that someday He would provide us all with a rationale for allowing such wanton evil to strike the innocent. In an effort to comfort the survivors, he quoted a verse in Isaiah.

> When thou passest through the waters, I *will be* with thee; and through the rivers, they shall not overflow thee: when thou walkest through the fire, thou shall not be burned; neither shall the flame kindle upon thee. (Is. 43:2)

It was a wonderful sermon, but the bomb that destroyed that building did burn the victims. So how can we reconcile the beautiful words of Isaiah with the horrible reality confronting us in Oklahoma City? The answer is that we can't! Like the scientists who cling to the erroneous paradigm of evolution, clergymen who believe in a personal God that watches over and cares for us must protect their belief even in the face of tragic evidence to the contrary. As a result, rather than question the religious paradigm of a personal God, Dr. Graham is forced to question the reality before him. *Why does God allow this to happen?* he asks. Unable or unwilling to answer that question, he pretends that there is no answer and hopes, as Darwin did, that a future revelation will provide an explanation for the obvious disparity. Many of the Jews that survived the Holocaust had a similar experience. Traumatized by the horror of the genocide perpetrated against them, they turned to their rabbis, who made obviously contrived excuses for God's failure to intervene. Even in the face of overwhelming contrary evidence, clergymen dare not question the paradigm because they foolishly equate questioning the doctrine of a personal God with questioning His existence. Allow me to assure the reader that when all of our unfounded beliefs are exposed, God will still stand.

The awful truth, deliberately ignored by both science and religion, is the fact that we do not know the truth. We know that we are living on a planet orbiting the sun, but we don't know who we are, where we came from, how we got here or where we go when we die. It is absolutely essential for our peace of mind that we know the answers to these very basic questions. Yet when we turn to the "experts" in science and religion, we get mixed messages; even the professionals seem confused. Compounding the problem is the fact that, having settled for their current doctrines, scientists and clergymen are no longer looking for the truth. Are these salient questions unanswerable, or is it just that our preconceived beliefs are keeping us from continuing the search? The answer is obvious.

It is clear to the dispassionate observer that the intellectual disagreement between science and religion is thwarting efforts to solve the mystery of the origins of man. Though this debate was quite heated in earlier times, recent years have seen these antagonists settle into an intellectual détente. Now they are ignoring each other and trying to pretend that both theories are right: that somehow God used evolution to create man. What nonsense! It is impossible for these diametrically opposed theories to both be right. At the very least, one of these "truths," if not both, has to be wrong. This conflict over the mystery of our existence is at the center of the chaos that has engulfed human civilization, and it must be settled!

The Ancient Texts: History or Myth?

The evidence needed to solve the mystery of our existence is readily available in the ancient texts, but the split between science and religion has made any discussion of these important resources taboo. Throughout this book you will see that highly educated experts are refusing to consider evidence merely because it is located in historical texts that have been censured by the early leaders of their particular cult. Scientists ignore the pagan texts and censure the Bible; Jews, the New Testament, while Christians deliberately burned all of the non-biblical writings they could find. Each of these cults holds to a unique paradigm and refuses to consider the contradictory evidence of their adversaries. We will see how this intellectual turf war is at the heart of our failure to use the historical texts to solve the great mysteries of both the ancient and modern worlds. A prime example of this problem is the case of Heinrich Schliemann.

Born in Germany in 1822, Schliemann traveled extensively as a young man and later settled in Athens, where he studied ancient Greek, developing a passion for Homer's *Iliad* and *Odyssey*. His studies led him to see these writings in a new light: not as myth, but as history chronicled in poetic form. In 1869 he put forth his hypothesis that Homer's writings were historical accounts of actual events, rather than the creations of a poet's fantasy. He further suggested that if archeologists were to use Homer's writings as a guide, they could easily find the lost city of Troy. These ideas contradicted the conventional wisdom in atheistic science and academia that had declared Homer's writings to be myth. As a result, Schliemann faced a storm of hostile criticism, and his theory was summarily dismissed by the archeologists as the delusions of an amateur.

Undaunted by this criticism and convinced of the historical importance of the texts, Schliemann used his personal wealth to fund an expedition in search of the "mythical" city. In 1870, using Homer as his guide, he began excavations in the Troad on the hill of Hissarlik in modern Turkey. Within a year he had uncovered nine superimposed settlements, including that of Homer's Troy. Schliemann's discovery sent a shock wave through archaeology. His hypothesis that the ancient writings and traditions were more than myths now had corroborating evidence.

What was the reaction of the scientific community to this amazing discovery? Did they celebrate it and endeavor to see if the other "mythical" texts might be used to guide archeological exploration? Quite to the contrary, they did not. These ancient texts, both biblical and pagan, contradict modern science on two fundamental points. They universally affirm the existence of gods, and they all teach that man is an alien, rather than an ape. These facts stand in direct opposition to the core tenets of the cult of ideological science; therefore, scholars don't dare give these texts even the slightest hint of credibility, lest the fallacy of their atheistic and evolutionary beliefs be exposed. To avoid losing face, they simply discounted Schliemann's system and discovery as beginner's luck and tried to ignore the awful truth that his work was a bona fide affirmation of the ancient texts. That blatant denial of reality thwarted objective research on the gods of the ancient world and reinforced the psychological bonds that enslave modern scientists to their atheistic myths.

While it is true that some archeologists have acquiesced and now use the ancient texts to locate new digs, for some strange reason they have carefully declined to answer the key question posed by Schliemann. Are these ancient writings and traditions reliable sources of historical fact or merely the mythical fantasies of the authors? I come down on the side of Schliemann. His work proves that a literal interpretation of these texts in a historical context is not only safe, but very revealing. They only become dangerous when, as in religion, we try to apply their ancient laws and customs to modern man. Some of these texts and traditions give graphic reports of our ancestors being transported to earth from distant galaxies. They also declare that the gods who brought us to earth interacted with man in the ancient world, providing him with the vast knowledge needed to build the great pyramids, temples and advanced civilizations of that world. Even more compelling is the fact that the scribes accurately prophesied a coming war of the gods that would end the world. In the next chapter, we will see that the ancient world did indeed come to a cataclysmic end precisely when the scribes said it would, leaving man alone on the earth in the dark ages of the early modern world. So you can see that these ancient texts are indeed an invaluable source of critical knowledge on both the ancient and modern worlds.

How, then, did these histories become embroiled in such

controversy, and why is it so difficult to reach agreement on the veracity of their content? The problem has its roots in the events that followed the collapse of the ancient world. The loss of access to the power and knowledge of the ancients left man in deep ignorance. The only repositories of limited knowledge left were the human descendants of the monarchs, and the scribes and priests of the ancient civilizations. These knowledgeable members of the aristocracy had been schooled in the ancient traditions. This knowledge gave them great power, and they used it to keep their fellow men in slavery. As these elders died off, this remnant of supernatural knowledge was lost, and the ancient theocracies slowly evolved into the religions of the modern world. During the dark ages, the hierarchies in these modern religions began to use the ancient texts as resources for insight into the origins of mankind, our relationship with the ancient gods and the dynamics of the earthly environment. These early intellectuals took the first faltering steps in human scientific research. In the West, Christianity became predominant, and all scientific exploration had to conform with its interpretation of the biblical texts. Attempts to explain life's mysteries by anyone other than the Church hierarchy were vehemently repressed. This repression was often violent, and many intellectuals, even though they professed a belief in God, were imprisoned or burned to death for unorthodox theorizing. With the passage of time, many of the Church's core beliefs, based largely on misinterpretations of the Bible, were repudiated by secular thinkers, and a fierce controversy ensued. These intellectuals evolved into a loosely-knit discipline which we now call secular science, and they began to challenge Church science. This new science allowed for free and open debate on all issues, except one: religion. In a backlash against Church censorship, they determined to refute the very foundational tenets of the Bible and were willing to embrace almost any position which could be used to embarrass the clergy. In 1859, when Darwin put forth his hypothesis on evolution, atheistic scientists were beside themselves with excitement at the prospect of destroying the core belief of their religious adversaries. Based on this unsubstantiated hypothesis, they declared the created alien paradigm of the Bible to be myth and set out to prove that man was a descendant of the primates. That declaration obligated scientists to take the position that all of the ancient writings were based on mythical beliefs. They portrayed them as accurate accounts of what primitive people believed about their "mythical gods," but worthless in a scientific search for truth. Their atheistic beliefs caused scientists to wrongly conclude that the ignorance of Church leaders in the dark ages was grounds for ascribing similar ignorance to the ancient scribes. That skewed their whole view of the ancient world, so they abandoned these historical texts and began to establish a totally separate, "scientific" view of man and his environment.

No decision was more devastating to our efforts to understand the origins and future of man. Thousands of years of history and wisdom were written off as myth to accommodate the beliefs of Darwin. Acceptance of his theories became a prerequisite for ordination to the position of scientist in this atheistic cult, leaving these important texts all but useless decorations on the bookshelves of academia. By spurning the ancient texts, scholars left them open to interpretation by fanatic religious zealots of every stripe, leading to the deaths of hundreds of millions of people in wars, holocausts, jihads and pogroms. If we are to end this carnage and discover the truth, scientists will have to begin to study the biblical history of ancient Israel with the same vigor and respect that they give to the hieroglyphics of the Egyptians and Mayans. It is very revealing that archeologists, who call a press conference when they manage to interpret the most insignificant pagan hieroglyphic, are, for the most part, biblically illiterate.

The acceptance or rejection of ideas without certain proof is the sign of a cult-like mentality; in the formulation of their atheistic paradigm, scientists have been guilty of both. Before being contradicted by Darwin, the extraterrestrial origin of man was accepted science throughout the advanced civilizations of the ancient world. Merely contradicting the ancient paradigm does not refute it; therefore, it was incumbent on scientists to disprove it on the merits prior to demanding that we replace it with the theory of evolution. Instead, they simply rejected thousands of years of ancient history and science and blindly endorsed Darwin's unsubstantiated hypothesis. That type of unquestioning support for Darwin continues to this day, forcing scientists to assume the defensive posture of a cult. I do not use that pejorative lightly. The dictionary definition of the word cult is "the zealous devotees of a person, idea or intellectual fad." Any organization requiring singular dedication to a person or a hypothesis while suppressing knowledge that contradicts its beliefs meets that criterion. Since science requires fanatic devotion to Darwin and unquestioning fidelity to the theories of atheism and evolution, it precisely fits that description. As further evidence, I submit that scientists have banned serious study of the Bible from their temples (universities). Is a scientific cult that bans books on God any more legitimate or objective than a religious cult that bans books on sex?

The backlash in science against the repressive power wielded by the early Christian Church has proven to be as big an impediment to a proper understanding of the ancient world as the Church itself. It is time to admit that neither side has solved these great mysteries and that their philosophical cold war leaves little hope that they will do so anytime soon. These scholars are no longer seeking a universal truth on the origins of man, but rather to reinforce their respective beliefs. Like men in glass houses who dare not throw stones, their positions are so fragile that scientists and clergymen have become loath to even use the word *truth*. I once had a discussion with a science teacher who happened to be Jewish. He challenged me with the old reliable, "Your truth, my truth: Never the twain shall meet" cop-out to avoid facing the fact that opposing philosophies cannot both be right. This man seemed to view truth as some kind of competition between schools of thought where we get to choose the theory that best suits our tastes. When I proved to him that truth is universal, he changed tactics and challenged my credentials. He mentioned that his uncle was a rabbi who, though he had studied the Torah all of his life, still would not dare to claim that he knew the truth. I responded that if a lifetime of study didn't reveal the truth, his uncle was probably in the wrong line of work. His statement brought to mind the apostle Paul's description of the theologians of his day, when he said that they were "ever learning and never able to come to the knowledge of the truth."[7] Two thousand years after Paul made that observation, these intellectuals are still studying and apparently need a little more time.

Professional scholars have been taught that an inquisitive mind is an intelligent mind; therefore, they seem to value the art of questioning more than the art of finding a solution. It has been my experience that, in a dialog, their questioning is more of a defensive debating tactic than an honest attempt to understand and explore potential solutions to these yet unsolved mysteries. Finally I asked this teacher if he had a time frame within which an average person like myself could expect that either his uncle or his scientific colleagues might be able to make a final disclosure of the truth they're seeking? Annoyed by my impertinence, he criticized my "narrow view" of the truth and said that a formal education would have taught me to question rigid solutions to complex problems. I guess he thought I had arrived at my innovative conclusions without questioning.

If I had proposed something to this teacher that he could have easily debunked with his scientific dogma, he would have been more than happy to correct my error, pat me on the head and send me on my way. However, when he was unable to quickly refute my position, he simply wrote me off as a Neanderthal to avoid a troubling dialogue. These experts have become so mesmerized by their own rhetoric that they see no value in the thoughts of anyone not accredited by their particular cult. Scientists and clergymen alike must begin to recognize the innate intelligence of all men and allow that unaccredited thought does not necessarily constitute unintelligent thought. The work of Schliemann has shown us that many great discoveries do not come from accredited scholars.

Often these professionals are too busy defending their orthodoxy to brainstorm accepted dogma for new ideas. When the truth on God and the origins of man is finally discovered, it will most likely be the result of serendipitous curiosity on the part of an outsider like Schliemann or the determined work of an accredited free thinker like Galileo. In either case, you can be sure it will be ridiculed and rejected by the experts.

I don't want to be too critical of my brothers and sisters in academia, but years of complex studies tend to make educated professionals seek complicated solutions to simple problems. My wife, the lovely and loyal Josephine, served twenty-five years as a cargo agent for Eastern Airlines. She found that when tracing missing freight, trained agents frequently started their search by considering the most complicated explanation for a problem. Josephine became a top agent by always double-checking for lost freight where it was supposed to be. Over ninety percent of the time, she discovered that a well-trained agent had failed to see the package in or near its proper place. Scientists and clergymen should take a cue from Josephine and double-check for the truth where it's supposed to be—in the historical writings of the ancient civilizations. It is there they will find, as did Heinrich Schliemann, that truth is simple.

The Unexplained Great Loss of Knowledge

Modern scientists are beginning to recognize that the civilizations of the ancient world were ruled by highly intelligent beings. Thousands of years before the modern era, the ancient Egyptians knew the precise size, shape and density of the planet, the speed of its rotation, the length of its orbit around the sun, the 26,000-year cycle of the equinoxes, the acceleration of gravity and the speed of light. Recent research has revealed that they utilized highly technical mathematics, geometry and astronomy in order to create a precise calendar and design the Great Pyramid to reflect the fundamental dynamics of both the earth and its solar system. Yet when Columbus set sail in the year A.D.1492, most people thought that the earth was flat, and scientists believed that it was located at the center of the universe. It seems inconceivable, but all of the advanced knowledge used by the ancients had vanished from the face of the earth, and mankind did not even begin to rediscover it until the 17th century. A great mystery surrounds this bizarre loss of scientific knowledge. It appears that in the year A.D.70 an enormous cataclysm occurred, and the highly intelligent beings who ruled those civilizations disappeared, leaving man alone on the earth in the dark ages of the early modern world. Simple logic forces us to conclude that a vital secret which holds the key to man's existence on planet Earth is hidden in the unexplained collapse of the ancient world.

SEEKING A NEW PARADIGM

In the next chapter I will propose that the mysterious fall of the gods and collapse of the ancient world holds the key to a universal truth. That cataclysm is the greatest unsolved mystery in the history of our planet; yet it is virtually ignored by scientists and clergy alike. Scientists avoid this extraordinary event because addressing it would require them to acknowledge the role of supernatural forces in the ancient world. The clergy sidestep it because the solution will reveal that the gods of that world are no longer dealing with man. My approach to solving the mystery will be simple: using the system Heinrich Schliemann used to discover Troy, I'll take the scribes and historians of the early civilizations at their word. By correlating biblical and pagan history with the recorded accounts of the fall of Jerusalem in A.D.70, I will reveal the incredible solution to the mysterious fall of the gods, collapse of the ancient world and subsequent rise of man to rule the planet. That solution holds the key to our existence and a wonderful truth more beneficial to mankind than anything suggested in the current beliefs of either science or religion.

11. The Fall Of the Gods

It is difficult to ignore the fact that the ancient civilizations were ruled by forces more powerful than anything modern man is capable of comprehending. The extraordinary ruins of these societies reveal that the engineering and mathematical skill used by the ancients was far beyond the prowess of primitive humans. In addition, the histories of these peoples universally attribute this advanced knowledge to the gods. Dissatisfied with that explanation, modern scholars seem determined to prove them wrong. They claim that these early civilizations merely fabricated stories of gods to satisfy man's innate need to worship a higher power. Nonetheless, all of the highly intelligent scribes of the ancient civilizations—bar none—declared that they were interacting with supernatural forces and referred to those powerful beings as gods. Not one of these highly sophisticated civilizations was ever found to be atheistic. Further, scientists cannot point to a single writer in the ancient world who even hints that the gods were anything but a factual reality, nor a single person who declared himself to be an atheist. Only in the modern world, where the absence of supernatural power has produced the conflicting dogmas of science and religion, do we find people denying the existence of God. Notwithstanding, a poll of any cross-section of people here in the modern world will find that over ninety percent think that mankind and the universe are the creations of an Almighty God. For scientists to demand that we ignore thousands of years of biblical and pagan history, disregard the tangible evidence of supernatural intervention in the ancient civilizations, suppress our natural instincts to believe in God and unquestioningly allow them to impose Darwin's debunked hypothesis on our children is the rhetoric of zealots. Can you imagine the outrage of these academics if one of the other philosophical cults were to insist that we teach its unfounded beliefs in the public schools as fact? It is time for these scientists to address the reality that atheism is totally out of step with the intuitive logic of the common man. I think they would be well advised to place as much value on the natural instincts of humans as they do on those of the Spotted Owl.

Rather than simply admit that they're wrong about the ancient gods, scientists are desperately trying to prop up the collapsing theory of evolution. As we have seen, they recently put forth the theory of punctuated equilibrium in order to excuse their inability to produce transitional fossils. This ludicrous position rivals an earlier attempt to support their belief that the massive pyramids and temples of the ancients were built by primitive humans. In that case they simply produced the unfounded hypothesis that these great works were created by men using ropes, wedges and wooden rollers. This theory not only ignored the great technical genius of the ancients, but their practical problems as well. For instance, the great monolithic stones used to build the astronomical calendar of Stonehenge in England weigh as much as 45 tons each, yet they were dressed and accurately aligned to reflect the movement of the stars. Built between 1,000 and 3,000 b.c. by the ancient Druids, Stonehenge is no pile of rocks. It consists of a complex arrangement of monolithic stones and some 56 pits[1] carefully placed in a series of concentric circles. Archeologist Gerald S. Hawkins theorized that this astronomical observatory was used as a virtual daily calendar to track the seasons and coordinate religious rituals with the time of the equinoxes.

Despite their enormous size, both the upright stones and lintels were worked with great precision to follow the gentle curve of the entire circle.

The largest of these trilithons is 22 feet tall with an additional 8 feet below grade level.

The 30 stones in the original outer sandstone ring were joined at the top by huge lintels which interlocked through a system of mortise and tenon and dovetailed at the ends to form a perfect circle.

The nearest source of the hard sandstone used in the outer circle of Stonehenge is 18 miles away in Marlborough Downs. The nearest source for the 60 blue stones used in the inner circles is the Preseli Mountains in southeast Wales. The only possible route to Stonehenge in the latter case would involve both land and water transportation. It is an affront to common sense for scientists to suggest that this incredible feat was accomplished using ropes, wedges and rollers, or that the astronomical accuracy of the complex was a serendipitous accident that occurred when primitive people cavemen created a religion in order to calm their fears. Druid "mythology" holds that ancient priests used supernatural powers to levitate these monoliths and float them to the site.

A similar mystery exists on one of the Caroline islands of Micronesia, where an ancient fortress was constructed using 400,000 basalt blocks weighing as much as ten tons each. There is no source of this stone on the island, and the blocks appear to have come from a quarry on the mainland. Here again, legend has it that a supernatural magician used his powers to float the blocks out to the island.[2] Even more incredible are the stone blocks used in the foundation of the ruins of Baalbek in Lebanon. These giants, roughly 15' x 60' x 20', are estimated to weigh as much as two thousand tons apiece.[3] Are we to assume that they too were manipulated with ropes, wedges and rollers?

Describing Caesarea, the city and harbor that King Herod built to honor the Roman Emperor, the great Jewish historian Josephus says:

> . . .for the case was this, that all the sea-shore between Dora and Joppa, in the middle, between which this city is situated, had no good haven, insomuch that every one that sailed from Phoenicia for Egypt was obliged to lie in the stormy sea, by reason of the south winds that threatened them; which wind, if it blew but a little fresh, such vast waves are raised, and dash upon the rocks, that upon their retreat, the sea is in a great ferment for a long way. But the king, by the expenses he was at, and the liberal disposal of them, overcame nature, and built a haven larger than was the Pyrecum [at Athens;] and in the inner retirements of the water he built another deep station [for the ships also.]
>
> Now, although the place where he built was greatly opposite to his purposes, yet did he so fully struggle with that difficulty, that the firmness of his building could not easily be conquered by the sea; and the beauty and the ornament of the works were such, as though he had not had any difficulty in the operation; for when he had measured out as large a space as we have before mentioned, he let down stones into twenty-fathom water, the greatest part of which were fifty feet in length, and nine in depth, and ten in breadth, and some still larger. But when the haven was filled up to that depth, he enlarged that wall which was thus already extant above the sea, till it was two hundred feet wide...in order to break the force of the waves, whence it was called Procumatia, or the first

> breaker of the waves; but the rest of the space was
> under a stone wall that ran round it. On this wall were
> large towers, the principal and most beautiful of which
> was called Drusium, from Drusus, who was son-in-law
> to Caesar. (*Wars of the Jews,* Bk.1, 21:5–6 p.453)

Conservative estimates place the weight of a stone 50' x 9' x 10' in the hundreds of tons. It is one thing for scientists to claim that slaves moved the huge stones of Baalbek and Stonehenge across dry land with ropes, wedges and rollers, but it is quite another to claim that this system was used to place similar stones on the ocean floor at Caesarea. How did Herod get these huge stones out into a turbulent sea and put them in place in 120 feet of water? These monoliths were not just dropped on the bottom; they were joined together to make a sea wall at least 126 feet high and 200 feet wide. This incredible feat of engineering is far beyond the capabilities of modern construction companies.

The description of this magnificent city and harbor belies the "scientific" theory that the great edifices of the ancient world were built by humans using primitive tools. In fact, the intensity with which scientists reject suggestions of a supernatural intervention is, in itself, very telling. Compounding their predicament is the fact that the ancient texts unanimously attribute this genius to the gods, forcing scientists to declare not just the texts, but the cities they mention to be myth. For years, science simply denied the existence of many of these cities; however, like the "fabled" city of Troy, Caesarea too was recently discovered.

Most people who have survived the educational systems of Western societies have been so brainwashed by the atheists in academia that they have great difficulty perceiving the supernatural aspects of the ancient civilizations. Compounding the problem is the fact that words fail to adequately convey the magnificence of these empires, and it is nearly impossible to capture the grandiose scale of their ruins on film. Notwithstanding, throughout this chapter I will present additional pictures of the ancient edifices and quote expert testimony on the scale of these works in order to discredit the scientific dogma that says they were built by primitive humans.

THE GREAT PYRAMID

A short distance from the city of Cairo is a huge rocky plateau that the ancient Egyptians leveled to within a fraction of an inch. Called the Giza plateau, it sits at the edge of the Libyan desert and serves as the foundation for the most famous pyramids of Egypt. The largest of these, the Great Pyramid, was created using an estimated 2,500,000 limestone and granite blocks. The core blocks, some weighing up to 70 tons, are arranged in 201 stepped tiers that rise nearly 500 feet to a flattened point which once supported a huge capstone.

We can only stand before this amazing structure in awe and wonderment. Our minds boggle as we attempt to understand how these ancient people were capable of such grandeur. In his unprecedented book, *Secrets of the Great Pyramid*, Peter Tompkins struggles with this and other puzzling questions. Tompkins' book presents a compilation of the modern scientific research done on the Great Pyramid at Giza from the time of its discovery to the present day. He tracks all of the different attempts to solve the mystery of this colossal structure and puts them forth in a compelling manner, leaving one with the distinct impression that the clash of ideologies between science and religion is preventing us from solving the mysteries of the ancient world. I have encapsulated Tompkins' account of the early attempts to investigate the Great Pyramid in order to illustrate how this rivalry has stifled research on this and a variety of other ancient mysteries.

The Great Pyramid of Cheops.

Children playing on core blocks reveal the enormous dimensions of this colossal structure.

Pyramid dwarfs two-story buildings and some local Egyptians.

IDEOLOGY AND THE GREAT PYRAMID

Among the first to attempt to unlock the secrets of the Great Pyramid was a man from London by the name of John Taylor. In 1850, investigating the Pyramid from the perspective of a mathematician, Taylor reached several conclusions. First, he determined that the ancient Egyptians had utilized the value of π in its construction. Second, he became convinced that the Egyptians were using a system of measurement based on a fraction of the earth's polar axis and the length of the solar year. Third, he postulated that the Pyramid was a scale representation of the earth, with the perimeter of the base representing the circumference at the equator and the height representing the distance from the earth's center to the poles. Taylor was a religious man and a student of the Bible. This led him to conclude that the advanced science utilized in the construction of the Pyramid was transmitted to the Egyptians through divine revelation. Though both Taylor and his theories were viciously attacked by the atheists of his day, his mathematical calculations were eventually proven to be totally accurate.

Just before he died in 1864, Taylor gained the support of Professor Charles Piazzi Smyth, the Astronomer Royal of Scotland. Smyth presented a paper supporting Taylor to the Royal Society of Edinburgh, but was treated no better than his predecessor. In an effort to verify Taylor's postulations, Smyth decided to launch an expedition to measure the Pyramid, but the Society refused to assist in funding the project, calling it religious fantasy. Undaunted, Smyth and his wife went to Egypt with a vast array of the latest surveying and measuring instruments. They set up living quarters in a nearby empty tomb and began the work of recording detailed measurements of the Pyramid.

Smyth's work confirmed Taylor's assertion that the Egyptians had utilized the value of π in the Pyramid's construction, and he was astonished to find that the builders had indeed used a system of measurement based on a fraction of the polar axis of the planet, called the pyramid inch. Dividing the perimeter of the base by this measurement produced the figure 365.24—representing the exact number of days in the solar year. Unable to explain this incredible achievement, Smyth, a religious man like Taylor, attributed it to the intervention of the gods. Using the biblical examples of how God had given certain men detailed instructions for the building of Noah's Ark and the Temple in Jerusalem, he suggested that the Great Pyramid was the product of a similar divine intervention. Smyth published his findings, but they were not well received. His attempt to use biblical history to explain the obvious supernatural aspects of the Pyramid reaped an onslaught of negative articles by atheistic colleagues which left him an outcast among his peers. Nevertheless, he went on to postulate that the height of the Pyramid in sacred inches was equal to the number of kilometers in the earth's solar orbit. This, too, would later prove to be true. Speaking of Smyth's hypothesis, Tompkins says:

> The idea was received by some with derision, by others with acrimonious opposition. One reviewer remarked that Smyth's book contained "more extraordinary hallucinations than had appeared in any other three volumes published during the past century." A friendly reviewer summed up reaction to the book saying "it evoked numerous illustrations of envy, hatred, malice, and much uncharitableness from vain, flippant, and unqualified writers, the author being scoffed at, traduced, worried and all but argued with, by opponents who only succeeded in proving their egotistic inefficiency to apprehend the truth." (Tompkins, p.93)

> Smyth's mixing of religious and prophetic conclusions with sound scientific discoveries caused his entire theory to be discarded by the skeptics. To this day, the lampooning persists. One modern writer on pyramidology still refers to Smyth as the world's "pyramidiot," and laments that "such a first-class mathematical brain should have wasted its energies in so unprofitable a field." (Tompkins, p.94)

It is true that Smyth was a brilliant mathematician. In addition, he had a keen insight because his position that the wisdom of the ancients was the result of a supernatural intervention will eventually prove to be true. Smyth was among a select few within academia who had actually visited the Pyramid and were, figuratively speaking, yelling at the top of their professional voices that empirical research did not support orthodox science on the ancients; notwithstanding, the hierarchy refused to listen and proceeded with their atheistic agenda.

The attempt to discover the true purpose of the Pyramid received another setback when Robert Menzies, a fellow Scot and biblical scholar, published his findings on the mystery. Menzies discovered that a chronology of Egyptian history had been incorporated into the physical construction of the Pyramid's system of passageways. He then hypothesized that the system also contained a prophetic revelation that corroborated biblical prophecy on the end of the world. This only intensified the rhetoric of the debate and widened the growing chasm between the religious intellectuals, who saw the Pyramid as scientific evidence of the intervention of supernatural forces in the ancient world, and the growing number of atheistic followers of Darwin, who saw only the hand of man.

The next attempt to solve the mystery came from Sir William Flinders Petrie, a professional surveyor who became fascinated with the work of Taylor and Smyth. Petrie conducted an extensive survey of the entire complex on the Giza plateau in 1880. His measurements confirmed the position of Taylor and Smyth that the Pyramid's designers had incorporated π, but he refuted Smyth's theory that the perimeter of the base reflected the exact number of days in the year. This news was well received by academia, and he was awarded many honors for his work. Though Petrie's findings were later reversed, for many years his mistake relegated the brilliant work of Taylor and Smyth to the limbo of contested science. Tompkins says that academia's rush to accept Petrie's erroneous conclusions caused great harm to research on the Pyramid.

> For years Smyth's painstaking measurements, carefully
> collected and illustrated in several large volumes (which
> went through several editions in his lifetime), were
> labeled by academicians so much "trash and fancy."
> (Tompkins, p.94)

This is the typical response of the scientific community to new ideas. They were quick to endorse Petrie's flawed conclusions affirming their preconceived beliefs, yet they simply ignored his discoveries that contradicted accepted dogma. This tragedy is especially ironic because Petrie's brilliant work surveying the Pyramid revealed a slight hollowing of the core masonry on each of its sides.[4] That discovery was eventually used to affirm Smyth's theory that the builders had indeed incorporated their knowledge of the solar system into the structure of the Pyramid.

One of the most interesting Egyptologists covered by Tompkins was a structural engineer from the north of England. In the late 1880s, David Davidson, an avowed agnostic, set out to discredit Robert Menzies' theory that the passage system was designed as a structural revelation of biblical prophecy. But such was not to be. Davidson soon discovered that if he extended Petrie's meticulous measurements of the core masonry to the outside casing stones, the length of the base achieved not only matched Smyth's theoretical length, but the length of the solar year to within four points of decimal. He further discovered that this hollowing of the sides made it possible to take three different measurements of the base, and that each of those measurements corresponds to one of the three unique systems used to observe and calculate the length of a year. By including this difference in his calculations, Davidson was able to prove that the perimeter of the base not only represented the length of the solar year, but the lengths of the sidereal and anomalistic years as well.

Davidson's conclusions were immediately rejected by atheistic scholars. Nevertheless, his writings revived interest in the Pyramid and created a whole new group of advocates for the work of Taylor and Smyth. Now, new details of the Pyramid were constantly

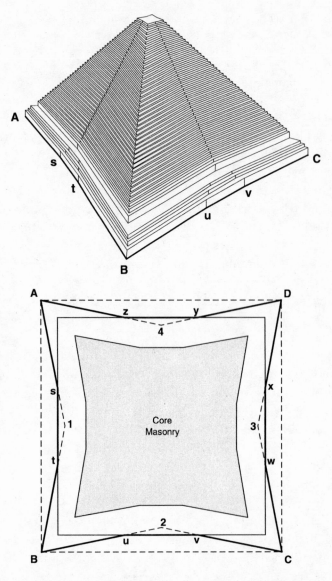

The solar year, which is obtained by observing the exact time between two successive equinoxes, is represented by ABCD in the diagrams above. The sidereal year, which is the time it takes a star to reappear in the same spot in the sky as seen from earth, is represented by AstBuvCwxDyz. The anomalistic, or orbital, year, which is the time it takes the earth to return to its closest point to the sun in its elliptical orbit, called perihelion, is represented by A1B2C3D4.

revealing the great genius of its design. It was discovered that the sum of the diagonals of the base of the Pyramid produced the extraordinary figure of 25,826.68 pyramid inches. This figure is roughly equal to the number of solar years in the *great year* of the precession of the equinoxes. The great year is caused by the slight wobbling of the earth on its axis and represents the number of years it takes for the axis of the earth to complete one circuit of this wobble.

His claim was challenged in the scientific community because the rate of precession is not constant, but Davidson solved the problem by using an average of the diagonals at various levels of the Pyramid to reach a mean length for the great year. Later, Morton Edgar, a Davidson supporter, found that the perimeter of the thirty-fifth course also provides this figure.

Davidson surmised that the Pyramid's designers must have been intimately acquainted with what he called "Natural Law,"[5] giving them highly advanced knowledge of the earth's dynamics. Awed by the fact that it had taken modern man more than two thousand years to begin to rediscover by experiment what the

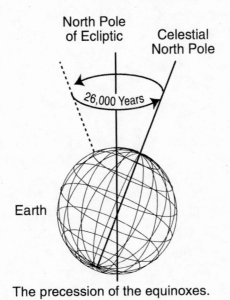

The precession of the equinoxes.

Egyptians knew at the outset, he concluded that they must have been communicating with supernatural beings. Davidson was so taken by the advanced nature of the intelligence behind the Pyramid's design that he was transformed from an agnostic to an advocate of Menzies' prophetic theories. Tompkins says of Davidson that he finally produced an encyclopedic work in support of Menzies and became convinced that the Pyramid was "an expression of the Truth in structural form" that would eventually prove the Bible to be the inspired word of God. This further antagonized his atheistic adversaries and accelerated the erosion of his position in academia.

When Davidson's sound mathematical evidence could no longer be denied, atheistic scholars argued that although the Pyramid's calibration was remarkably earth-commensurate, it was the result of coincidence, not design. This is a ludicrous position. If we just take the fact that this gigantic structure was built in perfect alignment with the four cardinal points of the compass and has remained constant for thousands of years, that alone makes their position untenable. In an effort to stem the tide of research revealing the supernatural implications inherent in the Pyramid, atheistic scientists began to concoct counter-proposals, suggesting ways in which humans could have built it using ropes, wedges and rollers. These reactionary theories were every bit as preposterous as the prophetic delusions of their biblical colleagues, yet because they affirmed atheistic dogma, they immediately gained wide acceptance in academia.

The work of Taylor, Smyth, Petrie, Menzies and Davidson exposes the key flaw in our efforts to solve the mysteries of the ancient world. When the research of these religiously oriented scientists revealed the supernatural aspects of the Pyramid, they naturally tried to link it to another mystical entity of that era: the Bible. Brainstorming from that perspective caused them to postulate a possible correlation between Pyramid chronology, biblical prophecy and the current events of the modern world. This was a tragic mistake that led some of the more radical supporters of these theories to create apocalyptic predictions that were really quite bizarre. The founder of the Jehovah's Witnesses, Charles Taze Russell, used a measurement of 1,914 sacred inches taken from the Pyramid's passageways to predict that the second coming of Messiah would end the world in 1914. Referring to another failed prophecy, Tompkins writes that,

> By 1920, when the waters of the Mediterranean failed to become thick and viscid, and the rivers and fountains of the world failed to turn to blood, as prophesied by Colonel J. Garnier on the basis of the Pyramid chronology, the whole subject became so unpleasant in the halls of academia that few professors dared mention the Pyramid as anything but the supposed resting place of the Pharaoh Cheops. (Tompkins, p.116)

This is a perfect illustration of how our failure to recognize the dichotomy between the ancient and modern worlds is thwarting scientific efforts to solve these mysteries. The Great Pyramid is the largest and most intricate edifice on planet Earth; yet, in order to avoid an "unpleasant" controversy between biblical and atheistic scholars over its meaning, scientists allowed this imponderable wonder to be reduced to a tomb. This inexcusable outcome is similar to the repercussions that followed the failure of these scholars to agree on the meaning of the ancient texts. In that case, the atheists simply declared these histories to be myth and walked away from the controversy. The Great Pyramid—a tomb? Biblical and pagan history—myth? Really, guys, can't we all just get along?

Academia is the only business where spoiled children can still get away with taking their ball and going home when they don't like the rules of the game. To suggest that this gigantic structure was built to hide a coffin rivals their equally ludicrous belief that it was built using ropes, wedges and rollers. Are they suggesting that the great Pharaoh of Egypt had his ministers build the largest structure on earth as some kind of gigantic mausoleum to avoid attracting grave-robbers?

Khafre Pyramid, Giza Plateau, Cairo. Some limestone casing stones remain near top.

I agree with biblical scientists who contend that the colossal structures of the ancient world are the work of supernatural beings and that because biblical history covers the same time span, it is reasonable to assume that there is a connection. However, biblical scientists only make themselves look foolish when they try to link these relics to modern man through contrived prophecy theories. On the other hand, the same can be said for the "ropes, wedges and rollers" crowd. When atheistic scientists try to downplay the supernatural nature of these colossal structures by concocting absurd human explanations for their construction, they, too, compromise the integrity of science and insult the intelligence of those they wish to persuade. These misguided attempts to intermingle ancient and modern realities are the key flaw in our efforts to solve the mysteries of the ancients. The current atheistic and biblical paradigms both require a continuing chronology between ancient and modern worlds. The clergy needs a continuum in order to claim that the gods of that world are still dealing with man, and the atheists need a continuum of evolution for Darwin to be right. Admitting the reality that an unbridgeable gap exists between the ancient and modern worlds will force them both to face the error of their current beliefs.

Scientific and religious hypotheses that try to bridge the unbridgable gap between these two utterly distinct worlds are inherently flawed. They reveal how the failure of academia to accept the totality of that dichotomy is thwarting their efforts to solve these mysteries. All academics, regardless of their ideological bent, must come to acknowledge the writings and relics of the ancient civilizations as important historical windows on the interaction between the gods and man in the ancient world and vigorously reject any attempt, from either side, to connect them to modern man. Only then can they begin to decipher the wonderful truth inherent in the collapse of the ancient world.

By the 1930s, the debate surrounding the Pyramid had become so ill-mannered that scientists sought ways to avoid the issue. Both sides had lost their ability to be objective and were more concerned with defending their beliefs than with seeking a solution to this great mystery. Each believed that the best defense was a strong offense, and the rhetoric became vicious. Theological scientists were called Bible-thumping Neanderthals, while the followers of Darwin where referred to as satanic atheists. In the end, they seemed to settle into a sort of ideological détente. Atheistic scientists ignored the factual evidence of a supernatural intervention implicit in the advanced science of the Pyramid, while their biblical colleagues continued to believe that their ludicrous prophecies would eventually come true. As a result, truth became an ideological choice rather than an intellectual pursuit, and "objective research" became a euphemism for seeking evidence that supported accepted dogma. With regard to this ideological warfare, Tompkins makes a revealing statement:

> In the conflict of opinions between biblical scholars
> and men of science, the true purpose of the Great
> Pyramid was buried in a rubble of verbiage.
> (Tompkins, p.107)

Over the years, almost nothing has changed. The true purpose of the Great Pyramid remains obscured by an avalanche of ideological rhetoric between the same two cults, one biblical and the other atheistic. The only difference is that the atheists have managed to win the propaganda war by tagging the biblical scientists with the stigma of failing to be objective. On that score, both sides fail to meet the mark. Professional pride has made it impossible for these experts to admit error, and there is no outside arbiter to judge their work; consequently, we are further from knowing the truth behind the ancient mysteries today than at any time in history.

To some degree, atheistic scientists are finally beginning to acknowledge that ancient Egypt was a center of highly advanced astronomy, physics, arithmetic, geometry, medicine, chemistry, geology, meteorology, geography and art. Peter Tompkins' book provides a wealth of evidence that the ancient Egyptians had utilized highly sophisticated mathematics in the construction of the Great Pyramid. He provides detailed evidence that the Egyptians designed it to depict the dynamics of both the earth and the solar system, and that each of its sides corresponds in perfect scale to one curved quarter of the northern hemisphere. Tompkins also makes the case that the Pyramid's precise alignment to the four corners of the compass make it an indestructible tool for surveying the earth and, further, that the accuracy of this alignment is such that modern compasses are adjusted to comply with its orientation. He draws the following conclusion.

> These cold facts should settle one whole facet of the mystery of the Great Pyramid. Clearly the ancient Egyptians knew the shape of the earth to a degree not confirmed till the eighteenth century when it was established that Newton was correct in his theory that the planet was somewhat flattened at the poles, and they knew the size of the earth to a degree not matched till the middle of the nineteenth century when it was measured with comparable accuracy by the German geodesist Friedrich Wilhelm Bessel. It is equally

evident that they could divide a day into 24 hours of 60 minutes and 60 seconds, and produce a unit of measure that was earth-commensurate just as Taylor, Smyth and Jomard had surmised. (Tompkins, p.211-12)

Whoever built the Great Pyramid, it is now quite clear, knew the precise circumference of the planet and the length of the year to several decimals—data which were not rediscovered till the seventeenth century. Its architects may well have known the mean length of the earth's orbit round the sun, the specific density of the planet, the 26,000-year cycle of the equinoxes, the acceleration of gravity and the speed of light. (Tompkins, p.xiv)

This is extraordinary knowledge! The ancient Egyptians not only knew that the earth was round, but they knew the exact speed of its rotation and could map and measure its surface with precise accuracy. Is there anyone, outside of the scientific community, who truly believes that the incorporation of all of this sophisticated knowledge into the Pyramid's construction was a mere coincidence? To the contrary, the Great Pyramid was the deliberate work of highly intelligent beings, advanced in all fields of science and in full control of their environment.

In harmony with this discovery, Erich von Daniken quotes[6] a Coptic manuscript from the Bodleiean Library, Oxford.[7] In it a scribe reports that an Egyptian king ordered his priests to deposit in the Great Pyramid all the knowledge they possessed of the heavenly spheres, including the fundamentals of geometry and mathematics. This was done to preserve the knowledge for those who had the ability to read the signs. The scribe's report led many researchers to think that there might be a storage room in the Pyramid that contained books or tablets inscribed with a wealth of ancient wisdom, but no such room was ever found. I think Tompkins' work reveals that this scribe's report is accurate, but that the priestly knowledge was incorporated into the Pyramid's design, making it an ingenious and almost indestructible record of ancient knowledge. I hope that someday all scientists will learn to read the signs, as did early Egyptologists like Taylor, Smyth, Menzies and Davidson.

These facts contradict the dogmatic beliefs of atheistic science that the ancient civilizations were merely the feeble attempts of primitive people learning to survive in a world that they could not understand. As we approach the 21st century, most people on the earth today are totally ignorant of the fact that the planet wobbles on its axis, creating the 26,000-year cycle of the equinoxes. Yet the Egyptians not only possessed this highly sophisticated knowledge, but incorporated it in the design of the Great Pyramid between five and ten thousand years ago. Equally astounding is the fact that at some point this advanced knowledge was lost, and we did not even begin to rediscover it until the 17th century.

A basic principle of science is that to solve a mystery one must ask the right question. In chapter seventeen of his book, Tompkins comes very close to asking this critical question, then backs away. The title of the chapter is "Decline of Ancient Knowledge," and it opens with the following all-important statement:

> What remains a mystery is how all the advanced
> science of the ancient Egyptians could have been lost
> for so many centuries. (Tompkins, p.214)

I agree! The loss of Egyptian science remains one of the greatest mysteries of planet Earth. The right question is, What caused the loss of that knowledge? Yet Tompkins treats this pivotal mystery as though it were a peripheral issue. He provides no explanation of his own, but quotes Professor Livio Catullo Stecchini,[8] who postulated that Alexander the Great had caused the loss of Egyptian geographic science when he destroyed Persepolis in the fourth century B.C. Even if Stecchini were right, this would only provide an explanation for the loss of some Egyptian science. How do we explain the loss of their ability to read their own writings (hieroglyphics)? Furthermore, what about all the other kingdoms and empires of the ancient world? The Mayans possessed sophisticated knowledge equal to, and in some areas surpassing, that of the Egyptians, and they, too, had an accurate calendar thousands of years before the modern era. What of the Incas, Romans, Greeks, Persians, Assyrians and Israelites? All of these civilizations were on a par with Egypt. How can we explain the great loss of knowledge, wealth and power manifest in the history and relics of these societies? This incredible decline was no local phenomenon; it occurred worldwide and had to have been precipitated by some catastrophic event. The scribes of these ancient civilizations recorded that they were interacting with gods who were providing them with sophisticated knowledge. The loss of that knowledge implies that, at some point, this communication was severed and man was left alone on the earth in the dark ages of the early modern world.

Atheistic scientists, evincing the same behavior they find so "unscientific" in their biblical counterparts, are sidestepping pivotal questions that might lead to uncomfortable answers. They are quick to supply volumes of minute details on the ancient civilizations as evidence of their expertise on the subject, but they carefully avoid addressing the bottom line. This academic preoccupation with the minutia of the ancients is a classic case of being incapable of seeing the forest for the trees. Tompkins' book is my most valued resource on the ancient Egyptians, yet even his pioneering work did not escape the intellectual limitations imposed by these academic blinders. His book provides 382 pages of detailed evidence that the ancient Egyptians possessed highly advanced knowledge in all areas of science, while just three pages are dedicated to presenting a half-hearted explanation for the mysterious disappearance of that knowledge. It is not essential for us to know the details of the advanced science used in the construction of the Great Pyramid, but it is critical that we understand how the Egyptians acquired that sophisticated intelligence so early in history and what caused it to suddenly disappear from the earth. Answering that question should have been the first order of business for all scientists. The fact that, two thousand years after the event, this pivotal mystery remains not just unsolved, but, for the most part, unstudied, is an indictment of their entire approach to the ancient world. In truth, they really don't want to know the answer to this philosophically loaded question because neither the question nor the answer fits the current paradigm. Atheistic science is based on Darwin's theory that man is on an ever-ascending evolutionary plane from primate to the sophisticated humans of the modern world. The precipitous loss of advanced science between the ancient and modern worlds leaves a gaping hole in that plane, thereby contradicting the very essence of the Darwinian model for man's existence.

Tompkins' decision not to take on this critical but academically incorrect mystery exposes the darker side of this problem, namely, the suppression of scientific freedom. In the introduction to the paperback edition of his book, he says:

> Dangerous as it is to venture into areas peripheral
> to accepted dogma, it is fortunate that even eight
> years after the original publication of this book, no
> one in or out of the academic world has faulted the
> basic premises.

I think that the fear expressed in this quote is very telling. All Tompkins did was document the highly advanced science used by the ancient Egyptians. Yet his reference to the danger in departing from accepted dogma indicates that he feared his work may have revealed more than was academically prudent to disclose. The scientific community has a low tolerance for unauthorized speculation on the supernatural nature of these phenomena, and Tompkins' words indicate that he knew his work had pressed the outer limits of orthodoxy. What right does the scientific community have to intimidate men like Tompkins? What are they afraid of? Why was it "dangerous" for Tompkins to address the supernatural aspects of the Great Pyramid, when they permeate every facet of his research? It is a sad commentary on the objectivity of this community when conceding the obvious requires an act of courage. Nevertheless, the mounting evidence that the ancient civilizations were ruled by highly advanced entities will soon force these atheists to address the name that the ancients used to describe those beings—*gods*.

Imagine for a moment that modern astronauts were to find a piece of space junk inscribed with hieroglyphics indicating that there was life on a distant planet. What do you think the reaction of academia would be if the clergy were to demand that we ignore the text because it didn't fit the Creation paradigm? Yet that is precisely the scientific response to the supernatural aspects of the ancient world. Scientists are quick to marvel at the great intellect of the Egyptians and Mayans when their hieroglyphics affirm the astronomy and mathematics of modern science, yet they totally discount the unequivocal testimony of the ancients that they received this knowledge from the gods. It is flagrantly unprofessional for scientists to disqualify the portions of these writings that contradict their atheistic beliefs. This selective science is a repeat of how they treated Schliemann's use of the Greek texts to discover Troy. They now admit that these ancient writings are a valuable resource in archaeology but continue to ignore their references to the gods. It is time for scientists to ask and answer the obvious questions: Where did the ancients get their advanced knowledge, What caused it to disappear from the earth and Who were the gods referred to in their writings?

In the 1970s, Erich von Daniken attempted to answer those very questions by forthrightly addressing the unexplained origin and sudden demise of the advanced knowledge used by the ancients. The histories of these civilizations indicate that this knowledge was jealously guarded by an elite caste of temple priests and members of the royal family, who refused to allow it to be shared with their lowly human subjects. Daniken was the first to consider the possibility that the members of this caste were extraterrestrials. He wrote several wonderful books suggesting that alien space travelers

The mysterious Great Sphinx of Giza is 187 feet long and 65 feet high with several rooms below grade level. This colossal sculpture, with its enigmatic stare and a veil of royalty draped over the back of its head, is cut from one gigantic boulder.

Hatshepsut's magnificent Temple at Deir El Bahri is built into the side of a rocky cliff.

had visited earth, created the great pyramids, temples and civilizations of the ancient world and then returned to their planets of origin, leaving mankind alone on the earth. Daniken was way ahead of his time. His departing aliens theory forced the scientific community to admit that they had no valid explanation for the sudden collapse of the ancient world. They rewarded his efforts to solve this great mystery with a storm of hostile criticism. In response, Daniken admitted that although his aliens theory could not be verified, he had put it forth in the hope of encouraging new thinking on the subject. I support him completely in that effort, and I agree with his contention that the highly advanced civilizations of the ancient world were the result of an extraterrestrial intervention. I also agree that the incredible loss of knowledge was caused by the hasty departure of those beings; however, the ensuing pages will show that I disagree with Daniken over the circumstances surrounding that world-ending event.

Why was Daniken vilified while Tompkins was given tacit approval? Unlike Tompkins, Daniken directly addressed the glaring reality that scientists were desperately trying to ignore. By formulating the first explanation for the early rise and sudden collapse of that world, he focused attention on this mystery and touched a raw nerve in the scientific community. He tried to soften the impact by replacing the term *gods* with *aliens*, but the scientists were not appeased. They knew that by using that term he was implying that these beings possessed powers beyond that of a normal human. Atheistic scientists saw Daniken's hypothesis as a direct challenge to both Darwin and their contrived human explanations for the supernatural achievements of the ancients. Consequently, they made every effort to discredit his work and continued to pretend that they had already solved these mysteries.

It is interesting to note that Daniken's books sold millions of copies, while Tompkins' work is largely unknown outside of academia. Tompkins' book stands alone as an entertaining and scholarly account of the advanced science utilized by the ancient Egyptians, but it failed to solve the great mystery of their existence.

By daring to directly address the supernatural aspects of the ancients, Daniken incurred the wrath of academia but engaged the imaginations of the masses. His aliens hypothesis reverberated in the intellect of the common man because it made sense. Most people instinctively know that the phenomenal structures of the ancient world required the intervention of supernatural forces, and the success of Daniken's books proves that the Darwinian atheists in academia have failed to convince them to the contrary. By using the term *aliens*, Daniken broke the chains of scientific and religious dogma shackling the imaginations of the hungry masses, allowing them (some for the first time) to dare to think for themselves. In an effort to stifle this "unprofessional" exuberance, scientists tried to discredit Daniken by incarcerating him in the literary prison of science fiction. There, he is in good company alongside other innovators and ostracized scholars like Galileo, Heinrich Schliemann, John Taylor, Charles Smyth, David Davidson and Michael Denton, to name just a few. Nevertheless, Daniken's theory of an alien intervention in the ancient civilizations is far closer to reality than the fictional science of atheistic academia, which says that primitive humans built those extraordinary cities with "ropes, wedges and rollers."

Even more incredible is the fact that scientists have also written off as mythology the entire body of work written by the scribes of the ancient Greeks, Egyptians, Romans, Mayans, Israelites and Persians. Some of the prestigious names in this group include Moses, Homer, Paul, Virgil and Josephus, but they were treated no better than Schliemann or Daniken by the atheists. All of these intellectual giants recorded that they were interacting with supernatural beings, providing descriptions of these gods[9] and demigods[10] that are simply incredible. For instance, it is reported that the demigod Incas, who controlled a large part of what is now South America, radiated such light that mortal men had to approach them backwards.[11] To look directly at the royal Inca was to die. Everywhere the Inca went, he was carried on a solid gold litter that weighed over 200 pounds, and he never set foot on the earth. The demigod King of Persia could not be approached without permission under pain of death, and everyone introduced into his presence had to touch their foreheads to the ground. In Egypt, Pharaoh's heralds preceded him warning "Earth, beware! Your god comes!" Anyone who touched the Pharaoh, even by mistake, was immediately put to death. In the biblical account of the confrontation between Pharaoh and the God of Abraham, when Moses turned his rod into a serpent as a sign of his authority, Pharaoh had his priests duplicate the deed. Pharaoh's arrogance led God to bring plagues on Egypt, culminating in the death of all the firstborn children in the empire, including that of Pharaoh himself. The power struggle ended when Moses parted the Red Sea so that the Israelites could cross to the other side and the great armies of Pharaoh were destroyed when the sea returned to its normal course. In ancient Israel, many great men of God possessed supernatural power. There was Joshua, who stopped the earth's rotation; Samson, who destroyed the amphitheater of the Philistines with his bare hands; David, who slew the giant Goliath with a single stone; and Jesus, who was called the "Son of God." It is recorded that Jesus provided evidence of his dominion over the forces of nature by walking on water, calming a raging sea, healing the sick and raising the dead. Having never known beings of this stature here in the modern world, we tend to question the credibility of these reports. Yet when we see scientists like Tompkins revealing that the writings of these ancient scribes evince a great knowledge of the universe and a high level of intellectual sophistication, we must temper our response.

HOMER & VIRGIL

Inspired by a heavenly spirit called a Muse, Homer and Virgil wrote fascinating accounts of the interaction between the pagan gods and man in the ancient world.[12] Their stories speak of supernatural beings who controlled the winds and seas, manipulating the elements in order to create violent storms. These gods also utilized giants and monsters to accomplish strange, diabolical schemes. For the longest time, academia treated these historians as fascinating writers who had a special penchant for creating mythical tales. This position totally ignored the fact that Homer, Virgil and their contemporaries all credited godly inspiration for the excellence of their work, not personal or even human intellect.[13] Unwilling to face that reality, atheistic academicians praised their genius for writing and simply ignored the declaration of the authors that their works were inspired by the gods. As we have seen, this position was debunked when Schliemann used Homer's histories to locate what science had called the "mythical" city of Troy. Since that time, many other cities that were mentioned only in the ancient writings have been found, making it increasingly difficult for scientists to deny the universal reports of a supernatural intervention in the ancient world. The outstanding question is, What caused the disappearance of the ancient gods and the collapse of their highly sophisticated civilizations? Modern scientists have suggested that the catastrophic demise of the ancient world was the result of climactic change, famine or the immorality of the ancients. These theories ignore the fact that the scribes clearly state that the gods of that world had dominion over nature. Such being true, we are obliged to conclude that only a force more powerful than the supernatural entities who ruled that world could have brought an end to their formidable empires.

The End Of The World

As extraordinary as the rapid advancement of the early civilizations may be, even more startling is their sudden decline. As I have mentioned previously, both the ancient Egyptians and

Mayans had an accurate celestial calendar thousands of years before the modern era. This proves that they had full knowledge of the solar system, including the shape of the earth, its speed of rotation and its course of revolution around the sun. However in 1492, when the explorer Christopher Columbus set sail, men lived in fear of falling off what they thought was a flat earth, which modern scientists believed was located at the center of the universe. This disparity of knowledge attests to the fact that at some point man's ability to communicate with the source of this advanced knowledge suddenly ceased. The questions we must ask are when and why this communication ended, and whom better to ask than the authors of the ancient texts themselves? Fortunately, they did provide us with the answers. Many of the biblical and pagan scribes recorded that the supernatural beings who visited earth in ancient times told men of a coming battle of the gods that would end the world as it then existed. The most explicit predictions of this coming cataclysm are found in the biblical history of ancient Israel.

ANCIENT ISRAEL

The Israelites were unique among the civilizations of the ancient world. Unlike the pagans, who worshiped many gods, Israel was monotheistic. It is also the only civilization from that world with a complete, detailed and cross-referenced history of its interaction with supernatural forces, as recorded in the Old and New Testaments of the Bible and the chronologies of Josephus. The Bible declares that our ancestors, Adam and Eve, were created in a place called the "Garden of Eden," where the god called Satan used a serpent[14] to hoodwink them into disobeying the instructions of the Creator.[15] This infraction caused God to exile Adam and Eve to Satan's dominion here on planet Earth. God promised our ancestors that He would eventually destroy the satanic forces that ruled the ancient world, making mankind the preeminent beings of the earth.[16]

The Bible records that to keep this promise, God chose the ancient Israelites and began to protect them from the satanic gods and demigods who ruled the pagan empires. With many great signs and wonders, He rescued them from slavery in Egypt and gave them the land of Israel. God promised to send the Israelites a Messiah[17] who would be born of a virgin and possess a godly spirit. It was prophesied that, in a great battle called Armageddon, this powerful demigod would destroy the satanic forces that ruled the ancient world and the meek (humans) would inherit the earth. History records that just such a battle took place in the skies over Jerusalem in the year A.D.70, which coincided with the fall of that magnificent city.

The destruction of Jerusalem is far more significant than the mere fall of an ancient metropolis. According to the Bible, the God of Abraham had designated that city as the place where He would dwell on earth. The heart of Jerusalem was the temple and the heart of the temple was a golden vessel, measuring approximately 4' x 3' x 3', called the Ark of the Covenant. Designed by God Himself, it was made of shit'tim wood, covered with pure gold both inside and out, and crowned with solid gold. On each corner there was a gold ring through which golden staves were placed to enable the Ark to be transported by the priests. God told Moses to place the tablets of the Covenant (written by God's own hand) in the Ark. The rod of Aaron[18] and a portion of God's manna were also kept with the Ark.[19] All of these things belonged to God and were evidence of His presence in Israel; therefore, they were both treasured by Him and sacred to the Israelites.

The Ark was more than a religious article; it was the physical manifestation of God's presence in Israel,[20] and it afforded the people many miracles.[21] The Ark even went with the Israelites when God caused them to be exiled in Babylon,[22] affirming His promise that one day they would return to Jerusalem.[23] The literal annihilation of Jerusalem and its temple, combined with the mysterious disappearance of the Ark, belies current Judeo-Christian hopes for a second restoration of Israel and provides undeniable evidence that the God of Abraham has completed His work among men and left the earth.

Why then do theologians contend that the God of Abraham is still dealing with man and that the prophecies of Armageddon are yet to be fulfilled? It is my position that modern-day religious leaders are repeating the mistakes of the biblical scientists who studied the Great Pyramid's passageways. By wrongly assuming that the prophecies of ancient Israel apply to modern man, theologians are forced to make ridiculous analogies to contemporary life. This has led to numerous deadlines for the end of the modern world, as well as accusations that individuals ranging from Adolph Hitler to Elvis Presley were the Antichrist.[24] My approach to the situation is simple. I consider all of the ancient texts to be historical accounts of events that took place in the ancient world. In that context they provide an easily discernible conclusion for the biblical and pagan prophecies and a logical explanation for the mysterious collapse of the ancient world.

According to Christianity, the biblical prophecies of Armageddon portray the end of the modern world. Contrary to this position, I contend that in these scriptures Jesus predicted the demise of the ancient world. With that premise as my foundation, I will now present the compelling correlation between the biblical prophecies of Armageddon and the fall of the gods, collapse of the ancient world and subsequent rise of man to rule the earth.

In order to understand the fall of Jerusalem, we must ask the right questions: Why did God allow the annihilation of the city and temple that He had declared to be His dwelling place on earth, and What are the implications of that cataclysmic event for modern man? In order to answer those key questions, it is essential to examine the events that preceded this catastrophe. Shortly before the destruction of Jerusalem, King Herod had convinced the high priests to allow him to totally dismantle God's temple in Jerusalem and rebuild the entire complex from scratch. He had spared no expense on the new edifice: the exterior was cut stone; the interior was cut from the cedars of Lebanon, and the roof was made of solid gold. However, unlike Ezra and Nehemiah's divinely ordained restoration of this holy place to honor the God of Abraham,[25] Josephus says that Herod's effort was an unauthorized attempt to leave an everlasting memorial to himself.

> And now Herod, in the eighteenth year of his reign, and after the acts already mentioned, undertook a very great work, that is, to build of himself the temple of God, and to make it larger in compass, and to raise it to a most magnificent altitude, as esteeming it to be the most glorious of all his actions, as it really was, to bring it to perfection, and this would be sufficient for an everlasting memorial of him.
> (*Antiquities of the Jews,* Bk.15, 11 p. 334)

Herod's unauthorized dismantling of God's temple to construct a memorial to himself was a desecration of the holy place. Jesus denounced this abomination, saying that they had turned his father's house into a "den of thieves."[26] In the New Testament book of Matthew, chapter 24, verses one to three, Jesus prophesies both the destruction of this temple and the end of the world.

> And Jesus went out, and departed from the temple: and his disciples came to him for to show him the buildings of the temple.
> And Jesus said unto them, See ye not all these things? verily I say unto you, There shall not be left here one stone upon another, that shall not be thrown down.

The temple was the center of the Jewish world, and when Messiah told his disciples that it would be destroyed, they were astounded. Notwithstanding, forty years later, Herod's everlasting memorial to himself was engulfed in a divine conflagration that consumed even the massive stones he had used in its construction.[27] The temple in Jerusalem was one of the largest and most magnificent wonders of the ancient world. The fact that not one stone of that structure can be found today is a stunning fulfillment of Jesus' prophecy.

Shocked by Messiah's prediction that their glorious temple would be destroyed, the disciples asked Jesus when this would happen and for a sign that would indicate the end was near.

> And as he sat upon the mount of Olives, the disciples
> came unto him privately, saying, Tell us, when shall
> these things be? and what the sign of thy coming, and
> of the end of the world?

It is important to note that when the disciples ask Jesus about the destruction of the temple, they link it to both his second coming and the end of the world. Since we know that the destruction of Herod's temple has already occurred, it is only logical to conclude that the second coming of Messiah and the prophesied end of the world coincided with that cataclysmic event in A.D. 70.

In verse 30 of Matthew 24, Messiah gives the disciples two signs that would portend the fulfillment of his prophecy.

> And then shall appear the sign of the Son of man in
> the heavens: and then shall all the tribes of the earth
> mourn, and they shall see the Son of man coming in
> the clouds of heaven with power and great glory.

It is essential to the proper interpretation of this verse that we identify the heavenly "sign of the Son of man" referred to by Jesus. This is easily determined because it was recorded in the Old Testament that a star in the heavens would be the sign of the coming of Messiah.[28] In the New Testament, historians recorded that this star stood over Bethlehem at the time of the birth of Jesus. With that in mind, I will quote from the historical account of the fall of Jerusalem by Josephus.

> . . .[The Jews in Jerusalem] did not attend, nor give
> credit, to the signs that were so evident, and did so
> plainly foretell their future desolation; but, like men
> infatuated, without either eyes to see or minds to
> consider, did not regard the denunciations that God

made to them. Thus there was a star resembling a sword, which stood over the city, and a comet, that continued a whole year. (*Wars of the Jews*, Bk.6, 5:3 p. 582)

In this historical record we find the exact fulfillment of Jesus' prophecy that in conjunction with the fall of Jerusalem, his star would appear in the heavens. The second sign in Matthew 24:30 makes the seemingly incredible assertion that men would actually see the Messiah coming in the clouds with power and great glory, but another quote from Josephus leaves no doubt as to the fulfillment of this prophecy as well.

. . .On the one-and-twentieth day of the month Artemisius, [Jyar,] a certain prodigious and incredible phenomenon appeared; I suppose the account of it would seem to be a fable, were it not related by those that saw it, and were not the events that followed it of so considerable a nature as to deserve such signals; for, before sun-setting, chariots and troops of soldiers in their armour were seen running about among the clouds, and surrounding [the] cities. Moreover, at that feast which we call Pentecost, as the priests were going by night into the inner [court of the] temple, as their custom was, to perform their sacred ministrations, they said that, in the first place, they felt a quaking, and heard a great noise, and after that they heard a sound as of a great multitude, saying, "Let us remove hence." (*Wars of the Jews,* Bk.6, 5:3 p. 582)

The Roman historian Tacitus, whose credentials are equivalent to those of Josephus, provides an equally graphic account of these events.

Prodigies had occurred, which [the Jews], prone to superstition, but hating all religious rites, did not deem it lawful to expiate by offering and sacrifice. There had been seen hosts joining battle in the skies, the fiery gleam of arms, the temple illuminated by a sudden radiance from the clouds. The doors of the inner

shrine were suddenly thrown open,[29] and a voice of
more than mortal tone was heard to cry that the gods
were departing. At the same instant there was a
mighty stir as of departure. (*History* 5:11–13)

These historical accounts of men in the clouds above Jerusalem cannot be taken lightly. Josephus and Tacitus are among the most esteemed chroniclers ever to record history, and their writings are considered to be above reproach. The fact that these reports precisely verify the fulfillment of Jesus' end of the world prophecies leaves one to wonder why they are not a cause for joyous celebration among biblical scholars.

There is one additional sign related in the end of the world prophecies. It is called the Rapture. This teaching promised the Messiah's followers that they would be supernaturally removed from the earth prior to the fall of Jerusalem. Many scriptures in both the Old and New Testaments relate this promise, but I will only quote one.[30] It is found in a letter from the apostle Paul to the disciples living in Thessalonica. In these verses, Paul not only reassured the disciples that they would be removed from the earth, but that their brethren who had died in faith would also be raised from the dead and taken out of the earth at the second coming of Messiah.

But I would not have you to be ignorant, brethren,
concerning them which are asleep, that you sorrow
not, even as others which have no hope.
For if we believe that Jesus died and rose again, even
so them also which sleep in Jesus will God bring with
him.
For this we say unto you by the word of the Lord, that
we which are alive and remain unto the coming of the
Lord shall not [precede] them which are asleep.
For the Lord himself shall descend from heaven with a
shout, with the voice of the archangel, and with the
trump of God: and the dead in Christ shall rise first:
Then we which are alive and remain shall be caught
up together with them in the clouds, to meet the Lord

in the air: and so shall we ever be with the Lord.
Wherefore comfort one another with these words.
(I Thes. 4:13–18)

Naturally, Christians claim that this prophecy is yet to be fulfilled, but the facts refute their position. First, Paul consistently uses the plural personal pronoun *we* in the text. When this pronoun is used, it includes the person writing; therefore, we must assume that Paul intended to participate in the Rapture. Second, he tells the disciples to comfort one another with his words. If they were not going to live to take part in this great event, Paul's promise would be a cruel hoax. Third, and most devastating to Christianity's claim that the apostles remained on the earth after the fall of the city, is the fact that all of the writers of the New Testament wrote extensively about the impending fall of Jerusalem, yet there is not one word in the Bible describing the incredible fulfillment of their prophecies. The only descriptions of this portentous event are found in the writings of Josephus and Tacitus. Because of the incredible significance of this cataclysm, these historians chronicled every detail of the destruction of this magnificent city and its glorious temple, including the supernatural phenomena that preceded its demise. If, as Christianity claims, the apostles remained on the earth after the city fell, why did they fail to mention the precise fulfillment of their own prophecies in their writings? The only possible explanation for such an egregious oversight is that they were not there to record it because, as promised by Paul, they had been supernaturally removed from the earth prior to the fall of the city. The importance of the fact that the New Testament does not contain a description of the fall of Jerusalem cannot be overstated. The absence of such a report time-locks its authorship in the period prior to A.D. 70, disqualifying Christianity's claim that God's dealings with man continued after the destruction of the temple. There is not one shred of historical evidence that the followers of Jesus were on the earth after A.D. 70. In addition, it is equally telling that the three definitive histories of ancient Israel, the Old and New Testaments of the Bible and the writings of Josephus, all terminate abruptly with the fall of Jerusalem.

In the face of the compelling biblical and historical evidence that this cataclysm was, in fact, the termination of God's dealings in Israel and the prophesied end of the world, one might ask how Christianity can continue to claim that these prophecies are yet to be fulfilled. This is achieved via the "suspended prophecy theory." This theory contends that some of the prophecies in Matthew 24 were fulfilled in A.D.70, and then God suspended the rest of the prophecies to allow the gospel to be taken to the heathens. Christians further assert that once God has allowed every heathen to hear the gospel, He will put all the entities found in Matthew 24 (the Roman Empire, Judea, the rebuilt temple etc.) back in place and fulfill the prophecies a second time. Since the creation of the modern state of Israel in 1948, Christianity has been full of absurd rumors that the Jews are about to rebuild the temple in preparation for Armageddon. I have been a serious student of biblical history for over twenty years. In all that time I have been unable to find any justification, biblical or otherwise, for the suspended prophecy theory. To the contrary, all of the historical evidence supports the position that these prophecies pertained to the ancient world. When Jesus prophesied that the temple would be destroyed, he specifically referred to the one built by Herod, which the disciples had shown him; he made no mention of another to be built in the future. Further, the expressed purpose of the temple was to house the Ark of the Covenant. Without the Ark, there is no need for Israel, or anyone else, to rebuild the temple.

Christianity's position also flies in the face of a statement by Jesus to his disciples as to who would see this great event.

> So likewise ye, when ye shall see all these things, know
> that it is near, even at the doors.
> Verily I say unto you. This generation shall not pass till
> all these things be fulfilled. (Mat. 24:33–34)

In verse 34 Jesus makes it crystal clear that all of his end of the world prophecies would be fulfilled in the same generation that witnessed the destruction of the temple. It is time for Judeo-Christian leaders to address the fact that all of the biblical prophecies that foretell the consummation of God's dealings with the ancient Israelites were fulfilled in the year A.D. 70.

Coming to grips with the reality that the Old and New Testaments of the Bible are a closed history of God's dealings in ancient Israel will enable us to be objective about the great cataclysm that is foretold in these writings. The end of the world predictions are inseparably linked to the destruction of Herod's temple and the second coming of Messiah. It is my contention that they were fulfilled in conjunction with the fall of Jerusalem in A.D. 70. That being true, then the rest of the prophecies which modern Christianity claims foretell the end of the modern world must also have referred to the ancient world. As evidence for that position, I will continue the quote of Jesus in Matthew, chapter 24 to verse seven and compare these further predictions with the historical account of the fall of Jerusalem as recorded by Josephus.

> And Jesus went out and departed from the temple: and his disciples came to him for to show him the buildings of the temple.
> And Jesus said unto them, See ye not all these things? verily I say unto you, There shall not be left here one stone upon another, that shall not be thrown down.
> And as he sat upon the mount of Olives, the disciples came unto him privately, saying, Tell us, when shall these things be? and what the sign of thy coming, and of the end of the world?
> And Jesus answered and said unto them, Take heed that no man deceive you.
> For many shall come in my name, saying, I am Christ; and shall deceive many.
> And ye shall hear of wars and rumors of wars: see that ye be not troubled: for all these things must come to pass, but the end is not yet.

> For [tribe] shall rise against [tribe], and kingdom
> against kingdom: and there shall be famines, and
> pestilences, and earthquakes in divers places.
> (Mat. 24:1–7)

For the sake of clarity, I have replaced the word *nation* with the word *tribe* in verse seven. At the time Jesus made this statement, there were no nations. The word *nation* did not come into use until the modern age. By translating the Greek word *ethnos* as *nation* in this verse, Christian translators caused many people to erroneously think that this prophecy was applicable to the modern world and that it would involve a worldwide conflict. That simply is not true. In this verse and others, Jesus is referring to a civil war among the twelve tribes of ancient Israel, with some tribes of the Northern kingdom fighting their brothers in the tribes of the Southern kingdom. A civil war cannot involve nations, therefore this verse must refer to the tribes and kingdoms of ancient Israel. Using the proper translation *tribe* instead of *nation* will relegate this prophecy to its proper time frame in the ancient world.[31]

Verses four through six in Matthew 24 are not definitive and could apply to any age. However, by tying the civil war described in verse seven to the destruction of the temple the disciples had shown him in verses one through three, Jesus forever links the fulfillment of this prophecy to the fall of Jerusalem in A.D.70. Jesus clearly states that in the end times a civil war among the Jews would engulf the city of Jerusalem, causing a great famine and pestilence, and culminating with the total destruction of the temple built by King Herod. This civil war is predicted throughout the Bible. In the Old Testament book of Micah, a prophecy reveals the terrible conditions that would exist in the city near the end.

> The good man is perished out of the earth: and there is
> none upright among men: they all lie in wait for blood;
> they hunt every man his brother with a net.
> That they may do evil with both hands earnestly, the
> prince asketh, and the judge asketh for a reward; and

the great man, he uttereth his mischievous desire: so
they wrap it up.
The best of them is a brier: the most upright is sharper
than a thorn hedge: the day of thy watchmen and thy
visitation cometh; now shall be their perplexity.
Trust ye not in a friend, put ye not confidence in a
guide: keep the doors of thy mouth from her that lieth
in thy bosom.
For the son dishonoreth the father, the daughter riseth
up against her mother, the daughter-in-law against her
mother-in-law; a man's enemies are the men of his own
house.

In that day also he shall come even to thee from
Assyria, and from the fortified cities, and from the
fortress even to the river, and from sea to sea, and from
mountain to mountain.
Notwithstanding the land shall be desolate that dwell
therein, for the fruit of their doings.
(Mic. 7:1–6, 12 & 13)

In the New Testament book of Luke, Jesus says:

For from henceforth there shall be five in one house
divided, three against two, and two against three.
The father shall be divided against the son, and the son
against the father; the mother against the daughter, and
the daughter against the mother; the mother-in-law
against the daughter-in-law, and the daughter-in-law
against her mother-in-law. (Luke 12:52–53)[32]

In these verses Jesus harmonizes with Micah, saying that in the
end times a civil war would turn brother against brother and father
against son. These prophecies are critical because in Josephus'
account of the fall of Jerusalem, he gives a precise record of their
fulfillment. Within forty years after these prophetic words were
spoken by Jesus, Josephus records that a vicious civil war swept
Jerusalem, forcing the Romans, who were responsible for its

orderly administration and protection, to restore order. Knowledge of this insurrection is essential to a proper understanding of these prophecies. Most people are under the false impression that the Romans destroyed Jerusalem, but Josephus reveals that it was the Jews themselves who caused its demise. His writings are crucial because they provide a detailed record of how a conflict between three different factions of zealots within the city forced the Romans to lay siege to Jerusalem in an attempt to stop the violence. By the time the Romans arrived, the rebels were engaged in open warfare, not just in the streets, but in the temple itself. They had already killed many of the high priests and were plundering the city and ravaging its people. As prophesied by Jesus, their reckless battles had destroyed the storehouses, causing a great famine and infestation of pestilence that was killing thousands of Jews daily. Josephus writes that these internal battles had left the streets of the city, the inner court of the temple and even the sanctuary littered with hundreds of thousands of dead bodies putrefying in lakes of blood. Many parts of the city were on fire and conditions were so terrible that most of the Jews longed for the Romans to intervene and bring an end to their suffering.[33]

> . . .many there were of the Jews that deserted every
> day, and fled away from the zealots, although their
> flight was very difficult, since they had guarded every
> passage out of the city, and slew every one that was
> caught at them, as taking it for granted they were
> going over to the Romans; yet did he that gave them
> money get clear off, while he that gave them none was
> voted a traitor. So the upshot was this, that the rich
> purchased their flight by money, while none but the
> poor were slain. Along all the roads also vast numbers
> of dead bodies lay in heaps, and even many of those
> that were so zealous in deserting, at length chose
> rather to perish within the city; for the hopes of burial
> made death in their own city appear of the two less
> terrible to them.[34] But these zealots came at last to that
> degree of barbarity, as not to bestow a burial either on

those slain in the city, or on those that lay along the roads; but as if they had made an agreement to cancel both the laws of their country and the laws of nature, and, at the same time that they defiled men with their wicked actions, they would pollute the Divinity itself also, they left the dead bodies to putrefy under the sun: and the same punishment was allotted to such as buried any, as to those that deserted, which was no other than death; while he that granted the favor of a grave to another, would presently stand in need of a grave himself. To say all in a word, no other gentle passion was so entirely lost among them as mercy; for what were the greatest objects of pity did most of all irritate these wretches, and they transferred their rage from the living to those that had been slain, and from the dead to the living. Nay, the terror was so very great that he who survived called them that were first dead happy, as being at rest already; as did those that were under torture in the prisons, declare, that, upon this comparison, those that lay unburied were the happiest. These men, therefore, trampled upon all the laws of man, and laughed at the laws of God; and for the oracles of the prophets, they ridiculed them as the tricks of jugglers; yet did these prophets foretell many things concerning [the rewards of] virtue, and [punishments of] vice, which when these zealots violated, they occasioned the fulfilling of those very prophecies belonging to their own country: for there was a certain ancient oracle of those men, that the city should then be taken and the sanctuary burnt, by right of war, when a sedition should invade the Jews, and their own hands should pollute the temple of God. Now, while these zealots did not [quite] disbelieve these predictions, they made themselves the instruments of their accomplishment.
(*Wars of the Jews*, Bk.4, 6:3, p. 536)

The role of the Romans with regard to the fall of Jerusalem has been grossly misstated. They did not come to Jerusalem as barbaric invaders seeking spoils. They came as the lawful administrators of a city under their protection. Josephus makes it very clear that God Himself had allowed the Romans to rule Jerusalem, in the same way that the Babylonians had been given dominion over the city on a previous occasion. It is important to understand that no foreign power could have controlled Jerusalem without the expressed permission of the God of Abraham.

The Romans became a great empire by relying heavily on the rule of law. When they engaged a people in battle, Roman commanders were required to adhere to the rules of war that had been set by the Senate. After submission, all peoples under their control were treated according to a strict code of military conduct. The Roman General Titus,[35] who commanded the legions surrounding the city, fully understood that his duty was to restore order with as little loss of life as possible. He had no instructions to change the status quo of Jerusalem or to alter the agreement that allowed the Jews to administer the city with Jewish law. Titus tried desperately to save Jerusalem and its magnificent temple, even going so far as to offer safe passage for the warring factions to continue their fight elsewhere or even to regroup and jointly fight the Romans if they would agree to leave the city. Unmoved by this show of goodwill, the tyrants refused all offers of clemency or pardon, and John, the leader of the strongest of the three factions involved in the civil war, sent a message out to Titus that he did not fear for Jerusalem because God Himself would defend both the city and His holy house.

> . . .The seditious cast reproaches upon [Titus] himself, and upon his father also, and cried out with a loud voice, that they contemned death, and did well to prefer it before slavery; that they would do all the mischief to the Romans they could while they had breath in them; and that for their own city, since they

were, as he said, to be destroyed, they had no concern
about it, and that the world itself was a better temple
to God than this. That yet this temple would be
preserved by him that inhabited therein, whom they
still had for their assistant in this war, and did
therefore laugh at all his threatenings, which would
come to nothing; because the conclusion of the whole
depended upon God only. These words were mixed
with reproaches, and with them they made a mighty
clamour. (*Wars of the Jews*, Bk. 5, 11:2 p. 565)

I will continue the chronology of Josephus as an exasperated
Titus begins his attack on the tower of Antonia.

And now Titus gave orders to his soldiers that were
with him to dig up the foundations of the tower of
Antonia, and make him a ready passage for his army
to come up; while he himself had Josephus brought to
him (for he had been informed that on that very day,
which was the seventeenth day of Panemus, [Tamuz,]
the sacrifice called "the Daily Sacrifice" had failed, and
had not been offered to God for want of men to offer
it,[36] and that the people were grievously troubled at it)
and commanded him to say the same things to John
that he had said before, that if he had any malicious
inclination for fighting, he might come out with as
many of his men as he pleased, in order to fight,
without the danger of destroying either his city or
temple; but that he desired he would not defile the
temple, nor thereby offend against God. That he
might, if he pleased, offer the sacrifices which were
now discontinued, by any of the Jews whom he should
pitch upon. Upon this, Josephus stood in such a place
where he might be heard, not by John only, but by
many more, and then declared to them what Caesar
had given him in charge, and this in the Hebrew
language. So he earnestly prayed them to spare their
own city, and to prevent that fire which was just ready
to seize upon the temple, and to offer their usual

sacrifices to God therein. At these words of his a great
sadness and silence were observed among the people.
But the tyrant himself cast many reproaches upon
Josephus, with imprecations besides; and at last added
this withal, that he did never fear the taking of the city,
because it was God's own city.
(*Wars of the Jews*, Bk. 6, 2:1 p. 574)

In answer to John's obscenity, Josephus rebuked him with
a loud voice:

To be sure, thou hast kept this city wonderfully pure
for God's sake! the temple also continues entirely
unpolluted! Nor hast thou been guilty of any impiety
against him, for whose assistance thou hopest! He still
receives his accustomed sacrifices! Vile wretch that
thou art! if anyone should deprive thee of thy daily
food, thou wouldst esteem him to be an enemy to thee;
but thou hopest to have that God for thy supporter in
this war whom thou hast deprived of his everlasting
worship! and thou imputest those sins to the Romans,
who to this very time take care to have our laws
observed, and almost compel these sacrifices to be still
offered to God, which have by thy means been
intermitted! Who is there can avoid groans and
lamentations at the amazing change that is made in
this city? since very foreigners and enemies do now
correct that impiety which thou hast occasioned: while
thou, who art a Jew, and was educated in our laws, art
become a greater enemy to them than the others! But
still, John, it is never dishonourable to repent, and
amend what hath been done amiss, even at the last
extremity. Thou hast an instance before thee in
Jechoniah, the king of the Jews, if thou hast a mind to
save the city, who, when the king of Babylon made
war against him, did, of his own accord, go out of this
city before it was taken, and did undergo a voluntary
captivity with his family, that the sanctuary might not
be delivered up to the enemy, and that he might not

see the house of God set on fire: on which account he is celebrated among all the Jews, in their sacred memorials, and his memory is become immortal, and will be conveyed fresh down to our posterity through all ages. This, John, is an excellent example in such a time of danger; and I dare venture to promise that the Romans shall still forgive thee. And take notice, that I, who make this exhortation to thee, am one of thine own nation; I, who am a Jew, do make this promise to thee. And it will become thee to consider who I am that gave thee this counsel, and whence I am derived[37]; for while I am alive I shall never be in such slavery as to forego my own kindred, or forget the laws of our forefathers. Thou hast indignation at me again, and makest a clamour at me, and reproachest me; indeed, I cannot deny but I am worthy of worse treatment than all this amounts to, because, in opposition to fate, I make this kind invitation to thee, and endeavor to force deliverance upon those whom God hath condemned. And who is there that does not know what the writings of the ancient prophets contain in them, —and particularly that oracle which is just now going to be fulfilled upon this miserable city?—for they foretold that this city should be then taken when somebody shall begin the slaughter of his own countrymen![38] and are not both the city and the entire temple now full of the dead bodies of your countrymen? It is God therefore, it is God himself who is bringing on this fire, to purge that city and temple by means of the Romans, and is going to pluck up this city, which is full of your pollutions.
(*Wars of the Jews*, Bk. 6, 2:1 p. 574)

This is a very important statement. Josephus was a high priest who was an expert in Jewish law and history, and he declares that the ancient prophecies are about to be fulfilled. He again harmonizes with Jesus and the prophets as he tells the seditious[39] zealots that God is going to use divine fire[40] to destroy the city to its very foundation stones.[41]

As Josephus spoke these words with groans, and tears
in his eyes, his voice was intercepted by sobs. However,
the Romans could not but pity the affliction he was
under, and wonder at his conduct. But for John, and
those that were with him, they were but the more
exasperated against the Romans on this account, and
were desirous to get Josephus also into their power: yet
did that discourse influence a great many of the better
sort; and truly some of them were so afraid of the
guards set by the seditious, that they tarried where they
were, but still were satisfied that both they and the city
were doomed to destruction. Some also there were
who, watching for a proper opportunity when they
might quietly get away, fled to the Romans, of whom
were the high priests Joseph and Jesus, and of the sons
of high priests three, whose father was Ishmael, who
was beheaded in Cyrene, and four sons of Matthias, as
also one son of the other Matthias, who ran away after
his father's death, and whose father was slain by
Simon, the son of Gioras, with three of his sons, as I
have already related: many also of the other nobility
went over to the Romans, together with the high
priests. Now Caesar not only received these men very
kindly in other respects, but, knowing they would not
willingly live after the customs of other nations, he sent
them to Gophna, and desired them to remain there for
the present, and told them, that when he was gotten
clear of this war, he would restore each of them to
their possessions again; so they cheerfully retired to
that small city which was allotted them, without fear
of any danger. But as they did not appear, the seditious
gave out again that these deserters were slain by the
Romans, —which was done in order to deter the rest
from running away by fear of the like treatment. This
trick of theirs succeeded now for a while, as did the
like trick before; for the rest were hereby deterred
from deserting, by fear of the like treatment.
(*Wars of the Jews*, Bk. 6, 2:2 p. 575)

Here we see the great respect and kindness with which Titus treated the Jews, against the treachery of the seditious. In order to discourage more Jews from fleeing the city, John claimed that Titus had killed these noblemen, striking fear in all who desired to do likewise. When Titus heard this, he brought those who had escaped back to the city and had them go around on the wall with Josephus so that the people could see his kindness to those who wanted peace. This encouraged a great many more Jews to go over to the Romans. These joined the others in reproaching the seditious for their impiety and urged them to save the temple. Unmoved, John loudly contradicted their statements and, in defiance, turned his engines to throwing great numbers of stones, arrows and javelins at the inner courts of the temple, which he did not yet control. Many of his own people were killed in this manner, and he further polluted the holy place by leaving their bodies to lie unburied, decomposing in the inner precincts of the temple itself. The seditious continually committed such profane acts until even the Romans began to reproach them for breaking Jewish laws and called on them to save the temple. Titus himself became so deeply affected by John's irreverence that he began to reproach him personally.

> Have not you, vile wretches that you are, by our permission, put up this partition-wall before your sanctuary? Have not you been allowed to put up the pillars thereto belonging, at due distances, and on it to engrave in Greek, and in your own letters, this prohibition, that no foreigner should go beyond that wall? Have we not given you leave to kill such as go beyond it, though he were a Roman? And what do you do now, you pernicious villains? Why do you trample upon dead bodies in this temple? and why do you pollute this holy house with the blood both of foreigners and Jews themselves? I appeal to the gods of my own country, and to every god that ever had any regard to this place, (for I do not suppose it to be now regarded by any of them;) I also appeal to my own

army, and to those Jews that are now with me, and
even to you yourselves, that I do not force you to defile
this your sanctuary; and if you will but change the
place whereon you will fight, no Roman shall either
come near your sanctuary, or offer any affront to it;
nay, I will endeavour to preserve you your holy house,
whether you will or not.
(*Wars of the Jews*, Bk. 6, 2:4 p. 575)

Here again we see the determined efforts of Titus to spare the temple, but the tyrant took this speech as a sign of weakness rather than goodwill, and he reproached the general. Seeing that they would not be reasonable, even in the worst of circumstances, Titus reluctantly renewed the war.

As Titus prepared his final attack, these demonic madmen continued their internal war, consuming the very resources needed to resist the coming siege. Then, as a stratagem in one of their battles with the Romans, they set a fire in the temple cloisters that began to encroach on the temple itself.[42] Even that terrible disaster did not deter them from their civil war or move them to join forces against the advancing Romans. In fact, when Titus ordered his troops to try to quench these fires, the tyrants ordered their men to attack them, thwarting Roman efforts to save the temple.[43] Titus made every effort to keep the holy house itself from burning, but Josephus describes this inferno as being of divine origin.[44] Near the end, he gives a truly frightening description of this horrific conflagration.

While the holy house was on fire, everything was
plundered that came to hand, and ten thousand of
those that were caught were slain; nor was there a
commiseration of any age, or any reverence of gravity;
but children, and old men, and profane persons, and
priests, were all slain in the same manner; so that this
war went round all sorts of men, and brought them to
destruction, and as well those that made supplication
for their lives, as those that defended themselves by
fighting. The flame was also carried a long way, and

made an echo, together with the groans of those that were slain; and because this hill was high, and the works at the temple were very great, one would have thought that the whole city had been on fire. Nor can anyone imagine anything either greater or more terrible than this noise; for there was at once a shout of the Roman legions, who were marching all together, and a sad clamour of the seditious, who were now surrounded with fire and sword. The people also that were left above were beaten back upon the enemy, and under a great consternation, and made sad moans at the calamity they were under; the multitude also that was in the city joined in this outcry with those that were upon the hill; and besides many of those that were worn away by the famine, and their mouths almost closed, when they saw the fire of the holy house, they exerted their utmost strength, and brake out into groans and outcries again; Perea did also return the echo, as well as the mountains round about, [the city,] and augmented the force of the entire noise. Yet was the misery itself more terrible than this disorder; for one would have thought that the hill itself, on which the temple stood, was seething-hot, as full of fire on every part of it, that the blood was larger in quantity than the fire, and those that were slain more in number than those that slew them. . . . (*Wars of the Jews* Bk. 6, 5:1 p. 581)

The noise, devastation and world-ending consequences of this holocaust impose a specter of the supernatural. This was not a normal fire that one might expect to occur in, say, the battle for Atlanta during the American Civil War. It is more akin to the fire described in the Armageddon prophecies of the Bible. King David says:

Our God shall come, and shall not keep silence; a fire shall devour before him, and it shall be very tempestuous round about him. (Ps. 50:3)

The prophet Micah says:

> For, behold, the Lord cometh forth out of his place,
> and will come down, and tread upon the high places
> of the earth.[45]
> And the mountains shall be molten under him, and
> the valleys shall be cleft, as wax before the fire, and
> as the waters that are poured down a steep place.
> (Mic. 1:3–4)

Micah's graphic description of a molten mountain and a valley that would split open (cleft) and run like liquefied wax seems incredible, but in the New Testament, the apostle Peter echoes his portrayal of what it would be like in "the day of the Lord."

> But the day of the Lord will come as a thief in the
> night; in which the heavens shall pass away with a
> great noise, and the elements shall melt with fervent
> heat, the earth also and the works that are therein shall
> be burned up. (II Pet. 3:10)

The departing heavens described in this verse refer to the heavenly forces that departed the temple. Those gods did indeed pass away with a great noise as recorded by both Josephus and Tacitus.[46] The melting elements are Herod's temple and the hill upon which it sat, which became molten and seethed with fire, as witnessed by Josephus. The massive stones of that temple, which totally vanished in this inferno, liquefied as God opened the earth, exposing the nether world below, and Satan and his minions were cast into that lake of fire.[47] Truly, this was the prophesied Armageddon.

I realize that many people who have been schooled in the Judeo-Christian prophecies are looking for a future battle of the gods that will destroy the entire earth. This primarily Christian teaching was based on a misinterpretation of the Bible. The truth is that all of the end of the world prophecies referred to the annihilation of Jerusalem. The graphic descriptions of these events as portrayed by

Josephus evince that reality—the great battle of supernatural forces in the heavens, the terrifying sound of the gods as they departed the sanctuary, his surreal description of the prodigious noise of the fire as the earth opened and the molten core of the planet devoured the very stones of Herod's temple—yet even these compelling descriptions seem incapable of conveying the magnitude of this apocalyptic holocaust. It is critical to understand that the temple in Jerusalem was built to house the Ark of the Covenant, which was the physical revelation of God's presence in Israel. It had scourged their enemies and killed anyone who touched it. They had used its supernatural powers to bring down the walls of Jericho, part the waters of the Jordan and destroy the god of the Philistines.[48] The leaders of the civil war in Jerusalem knew that history, and a fierce battle was fought for control of the holy house. As we have seen, John, who had captured the temple, arrogantly warned Titus that God Himself would protect it and give him victory over the Romans; yet the holy house was totally destroyed and the Ark vanished.[49]

Where is the powerful Ark of the Covenant that revealed God's presence on earth? It is indeed gone; however, despite the entertaining movie portraying the exploits of Indiana Jones,[50] I can assure you that God's Ark is not lost. Scriptures in both the Old and New Testaments of the Bible state that at the end of the world, the Ark would be removed from the earth.[51] Its absence, combined with the annihilation of the temple and Jerusalem, sends a disturbing, but unmistakable, message to all of mankind. It is time to face the reality that the God of Abraham has withdrawn His presence from the earth.

In the face of the compelling biblical and historical evidence that this cataclysm was, in fact, the termination of God's dealings in Israel and the prophesied end of the world, one might ask how Judaism can continue to claim that these prophecies are yet to be fulfilled. I put that very question to a member of the Union of Orthodox Rabbis in New York City and was surprised to find that, like the Christians, they too have a suspended prophecy theory. The rabbi I spoke with claimed that at the time the temple was destroyed, one of the high priests hid the Ark and that when the Messiah comes, he will rebuild the temple, remove the Ark from its hiding place and restore the Jews to Jerusalem. In light of the fact that every detail of what would happen to the Jews was always prophesied beforehand, I asked him if a prophecy in the Old Testament foretold the hiding of the Ark; he answered no. I asked if he could tell me the name of the high priest who had hidden the Ark; he again responded in the negative. I then asked if a record of this concealment were documented in any of the historical accounts of the fall of Jerusalem, and he again said no. At that point I could see that there was no sense in pursuing the matter, so I thanked the rabbi for his time and ended the conversation.

Like the Christians, who have neither biblical justification for their suspended prophecy theory nor evidence that the followers of Jesus were on the earth after the fall of Jerusalem, the Jews cannot point to a single prophecy which supports their claim that God instructed the high priests to hide the Ark, nor to any record of it actually being secreted away by the priests when the temple was destroyed. It appears that the story of the Ark being hidden was simply made up out of whole cloth to avoid having to face the inescapable reality evinced by its absence.

As the Romans proceeded with the siege of Jerusalem, the demon-possessed leaders of the civil war continued fighting both each other and the advancing troops until they were taken by force. By that time, the apocalyptic fire that had literally consumed the temple had also reduced the city to a smoldering ruin. As the battle neared the end, the following quote from Josephus reveals the demeanor of both the Romans and the seditious rebels in Jerusalem.

> And now the Romans, upon the flight of the seditious
> into the city, and upon the burning of the holy house
> itself, and of all the buildings lying round about it,
> brought their ensigns to the temple, and set them over
> against its eastern gate; and there did they offer
> sacrifices to them, and there did they make Titus
> imperator, with the greatest acclamations of joy. And

now all the soldiers had such vast quantities of the
spoils which they had gotten by plunder, that in Syria a
pound weight of gold was sold for half its former value.

With the seditious tyrants forced out of their stronghold in the
temple by the fire and the advancing Roman legions, their demeanor is
less ferocious and they ask to discourse with Titus. Here again Josephus
reveals the genuine desire of the Romans to spare the city as the general
makes them yet one more offer to save what was left of it.

But as for the tyrants themselves, and those that were
with them, when they found that they were
encompassed on every side, and, as it were, walled
round, without any method of escaping, they desired to
treat with Titus by word of mouth. Accordingly, such
was the kindness of his nature, and his desire of
preserving the city from destruction, joined to the
advice of his friends, who now thought the robbers
were come to a temper, that he placed himself on the
other side of the outer [court of the] temple; for there
were gates on that side above the Xystus, and a bridge
that connected the upper city to the temple. This
bridge it was that lay between the tyrants and Caesar,
and parted them; while the multitude stood on each
side, —those of the Jewish nation about Simon and
John, with great hope of pardon; and the Romans
about Caesar, in great expectation how Titus would
receive their supplication.·

So Titus charged his soldiers to restrain their rage, and
let their darts alone, and appointed an interpreter
between them, which was a sign that he was the
conqueror, and first began the discourse, and said, "I
hope you, sirs, are now satisfied with the miseries of
your country, who have not had any just notions either
of our great power or of your own great weakness; but
have, like madmen, after a violent and inconsiderate
manner, made such attempts as have brought your
people, your city, and your holy house to destruction.

You have been the men that have never left off
rebelling since Pompey first conquered you; and have,
since that time, made open war with the Romans.
Have you depended on your multitude, while a very
small part of the Roman soldiery have been strong
enough for you? Have you relied on the fidelity of
your confederates? and what nations are there, out of
the limits of our dominion, that would choose to assist
the Jews before the Romans? Are your bodies stronger
than ours? nay, you know that the [strong] Germans
themselves are our servants. Have you stronger walls
than we have? Pray, what greater obstacle is there than
the wall of the ocean, with which the Brittons are
encompassed, and yet do adore the arms of the
Romans? Do you exceed us in courage of soul, and in
the sagacity of your commanders? Nay, indeed, you
cannot but know that the very Carthaginians have
been conquered by us. It can therefore be nothing
certainly but the kindness of us Romans which hath
excited you against us; who, in the first place, have
given you this land to possess; and, in the next place,
have set over you kings of your own nation; and, in the
third place, have preserved the laws of your forefathers
to you, and have withal permitted you to live, either by
yourselves or among others, as it should please you;
and, what is our chief favour of all, we have given you
leave to gather up that tribute which is paid to God,
with such other gifts that are dedicated to him; nor
have we called those that carried these donations to
account, nor prohibited them; till at length you became
richer than we ourselves, even when you were our
enemies; and you made preparations for war against us
with our own money: nay, after all, when you were in
the enjoyment of all these advantages, you turned your
too great plenty against those that gave it you, and,
like merciless serpents, have thrown out your poison
against those that treated you kindly. I suppose,
therefore, that you might despise the slothfulness of
Nero, and, like limbs of the body that are broken or
dislocated, you did then lie quiet, waiting for some

other time, though still with a malicious intention, and have now shown your distemper to be greater than ever, and have extended your desires as far as your impudent and immense hopes would enable you to do it. At this time my father came into this country, not with a design to punish you for what you had done under Cestius, but to admonish you; for, had he come to overthrow your nation, he [would have] run directly to your fountainhead, and immediately laid this city waste; whereas, he went and burnt Galilee, and the neighbouring parts, and thereby gave you time for repentance; which instance of humanity you took for an argument of his weakness, and nourished up your impudence by our mildness. When Nero was gone out of the world, you did as the wickedest wretches would have done, and encouraged yourselves to act against us by our civil dissensions, and abused that time when both I and my father were gone away to Egypt, to make preparations for this war. Nor were you ashamed to raise disturbances against us when we were made emperors, and this while you had experienced how mild we had been, when we were no more than generals of the army; but when the government was devolved upon us, and all other people did thereupon lie quiet, and even foreign nations sent embassies and congratulated our access to the government, then did you Jews show yourselves to be our enemies. You sent embassies to those of your nation that are beyond Euphrates, to assist you in your raising disturbances; new walls were built by you round your city, seditions arose, and one tyrant contended against another, and a civil war broke out among you; such, indeed, as became none but so wicked a people as you are. I then came to this city, as unwillingly sent by my father, and received melancholy injunctions from him. When I heard that the people were disposed to peace, I rejoiced at it: I exhorted you to leave off these proceedings before I began this war; I spared you even when you had fought against me a great while; I gave my right hand as security to the deserters; I observed what I had

promised faithfully. When they fled to me, I had
compassion on many of those that I had taken captive;
I tortured those that were eager for war, in order to
restrain them. It was unwillingly that I brought my
engines of war against your walls; I always prohibited
my soldiers, when they were set upon your slaughter,
from their severity against you. After every victory I
persuaded you to peace, as though I had been myself
conquered. When I came near your temple I again
departed from the laws of war, and exhorted you to
spare your own sanctuary, and to preserve your holy
house to yourselves. I allowed you a quiet exit out of it,
and security for your preservation: nay, if you had a
mind, I gave you leave to fight in another place. Yet
have you still despised every one of my proposals, and
have set fire to your holy house with your own hands.

And now, vile wretches, do you desire to treat me by
word of mouth? To what purpose is it that you would
save such a holy house as this was, which is now
destroyed? What preservation can you now desire after
the destruction of your temple? Yet do you stand still
at this very time in your armour; nor can you bring
yourselves so much as to pretend to be suppliants even
in this your utmost extremity! O miserable creatures!
what is it you depend on? Are not your people dead? is
not your holy house gone? is not your city in my
power? and are not your own very lives in my hands?
And do you still deem it a part of valour to die?
However, I will not imitate your madness. If you
throw down your arms, and deliver up your bodies to
me, I grant you your lives; and I will act like a mild
master of a family; what cannot be healed shall be
punished, and the rest I will preserve for my own use."

To that offer of Titus they made this reply: —That
they could not accept of it, because they had sworn
never to do so; but they desired they might have leave
to go through the wall that had been made about them,
with their wives and children; for that they would go

into the desert, and leave the city to him. At this Titus
had great indignation; that, when they were in the case
of men already taken captives, they should pretend to
make their own terms with him, as if they had been
conquerors! So he ordered this proclamation to be
made to them, That they should no more come out to
him as deserters, nor hope for any further security; for
that he would henceforth spare nobody, but fight them
with his whole army; and that they must save
themselves as well as they could; for that he would
from henceforth treat them according to the laws of
war. So he gave orders to the soldiers both to burn and
plunder the city; who did nothing indeed that day; but
on the next day they set fire to the repository of the
archives. . . . (*Wars of the Jews* Bk. 6, 6:1-3 p. 583)

This was a tragic loss. The repository of the archives held the
recorded lineages of the Jews, without which it would never again
be possible to determine who was and who was not a Jew. Unlike
the Babylonian destruction of the city, where God made specific
provisions for the Jews to maintain their identity in captivity, in
A.D. 70 He deliberately allowed the lineages to be destroyed. The
loss of these records is an unmistakable message from God that He
was ending His work in Israel.

Josephus' writings provide clear evidence as to who was
responsible for the destruction of Jerusalem—as prophesied, it was
the demon-possessed leaders of the civil war. In summary he writes:

. . .for I venture to affirm, that the sedition destroyed
the city, and the Romans destroyed the sedition, which
it was a much harder thing to do than to destroy the
walls. . . (*Wars of the Jews* Bk. 5, 6:1 p. 557)

When it was over, Josephus estimated that more than 1,100,000
Jews had been killed and an additional 97,000 had been taken
captive as slaves. He further states that most of the Jews were
consumed by the rampant famine and pestilence that ravaged the

city as a result of the civil war, an exact fulfillment of the prophecies of both Micah and Jesus.

These historical accounts of the fall of Jerusalem testify to the fact that, precisely as prophesied in the ancient writings, God Himself allowed the seditious Jews to bring this cataclysm on themselves in the end times. They also reveal that, in conjunction with the fall of the city, a great battle of the gods was engaged in the heavens. In that battle, the satanic forces that had ruled the ancient civilizations were destroyed by the second coming of Messiah, leaving mankind alone on the earth.

Satan Is Dead

Identifying Satan as the supernatural power and intelligence behind the ancient civilizations is a prerequisite to solving the mystery of their demise. However, as the refusal of atheistic scientists to utilize the writings of Homer kept them from discovering the "mythical" city of Troy, their rejection of the Bible and its "mythical" Satan thwarts their efforts to understand the collapse of that world. Allow me to assure you that, as the historical writings of Homer verified the existence of Troy, correlating the biblical texts with the pagan will prove that Satan was the supernatural entity behind the advanced civilizations of that age. In addition, for those readers whom modern religion has traumatized through fear of this entity, fear not, because the evidence will also show that he has been dead for more than 1,900 years.

Judeo-Christianity's ludicrous portrayal of Satan as a bizarre spiritual creature who enjoys watching little boys masturbate, combined with the cult of academia's fanatical rejection of biblical history, has made it all but impossible to have a rational discussion of this being. Nevertheless, Satan is a legitimate historical figure and an indispensable component of any research on the ancient world. Biblical history records that as punishment for his rebellion against the

Almighty God and his corruption of Adam and Eve in the Garden of Eden,[52] God told Satan that He would destroy him and give mankind his place as ruler of the earth.[53] Knowledge of that prediction is essential to any scientific effort to explain the mysterious fall of the gods and collapse of the ancient world.

In the New Testament there is an incident that identifies Satan as the supernatural entity behind the powerful empires of the ancient world. In the year A.D.30, realizing that his prophesied day of reckoning with the Almighty God was near, Satan tried to bribe Jesus in an effort to entice him to join his rebellion. In the process, he made a highly revealing proposition.

> And the devil, taking [Jesus] up into a high mountain,
> showed unto him all the kingdoms of the world in [that]
> moment of time.
> And the devil said unto him, All this power will I give
> thee, and the glory of them: for that is delivered unto me;
> and to whomsoever I will give it.
> If thou therefore wilt worship me, all shall be thine.
> And Jesus answered and said unto him, Get thee behind
> me, Satan: for it is written, Thou shalt worship the Lord
> thy God, and him only shalt thou serve. (Luke 4:5-8)

There are two very important aspects of these verses. First, Satan shows Jesus all the kingdoms of the ancient world and offers to give them to him. These were the powerful and glorious empires of Egypt, Rome, the Incas and Mayans, Assyria, Greece and so on. It is important to note that in verse six Satan claimed to have dominion over those civilizations, and it is equally telling that Jesus didn't challenge his claim. Second, we must ask ourselves why a powerful being like Satan would feel constrained to lay all of the glorious wealth and power of the ancient world at the feet of a lowly Jewish carpenter. It is not unreasonable to conclude that Satan knew that Jesus was the promised Messiah and was desperately trying to make a deal to stave off his rapidly approaching confrontation with God. Satan's fears were not unfounded; the Bible is replete with prophecies that the Messiah of Israel would destroy him. The apostle Paul made one of these predictions to the disciples at Rome.

> And the God of peace shall crush Satan under your
> feet shortly. (Rom. 16:20)

The Greek word translated *shortly* in this verse describes a very limited time span. By using it, Paul deliberately indicated that this event was imminent, and history affirms the accuracy of his prediction. As we have seen, Josephus and Tacitus recorded that in A.D. 70 a battle between extraterrestrial forces raged in the skies over Jerusalem and a supernatural voice declared the departure of the gods. History shows that from that point forward, man's ability to communicate with the ancient gods was severed, and we found ourselves alone on the earth, heirs to the legacy of the gods.

> Blessed are the meek: for they shall inherit the earth.[54]
> (Mat. 5:5)

In this verse from the Sermon on the Mount, Jesus foretold the rise of mankind to rule the earth. This prophecy is very significant because the ruling class in the ancient civilizations consisted of gods that ruled from the heavens and their earthbound counterparts— the demigod Pharaohs, Emperors, Royal Incas, Mayan Chieftains and Kings. In addition, there were the descendants of the monarchs who, by birthright, automatically became members of these ancient aristocracies. Humans were mere slaves to these blue-bloods, and as such, they could be worked to death on meager rations, thrown to hungry lions for sport or ordered to participate in mortal combat with a fellow human in the amphitheaters of the gods. When Jesus prophesied that the meek would inherit the earth, it was inconceivable that man would ever replace the gods; but who today would deny the fulfillment of his incredible prophecy?

Though Messiah's prediction that God would give man the earth seems incredible, it is identical to King David's prophecy in the Old Testament book of Psalms:

> For [Satan's minions] shall be cut off: but those that
> wait upon the Lord, they shall inherit the earth.
> For yet a little while, and [Satan] shall not be: yea,
> thou shalt diligently [seek] his place, and it shall not
> be [found].
> But the meek shall inherit the earth; and shall delight
> themselves in the abundance of peace.
>
> I have seen [Satan] in great power, and spreading
> himself like a green bay tree.
> Yet he passed away, and, lo, he was not: yea, I sought
> him, but he could not be found. (Ps. 37:9–11, 35 & 36)

In this psalm, written almost a thousand years earlier, David harmonizes perfectly with Jesus by predicting that in the end times, the evil forces that ruled the ancient world would be destroyed and humans would become the preeminent beings of the earth. Who can deny the evidence that these prophecies were fulfilled? Exactly as King David predicted, the mighty empires of the ancients have collapsed, and humans now rule the earth.

In the New Testament, the apostle John makes a definitive statement as to the purpose for the coming of Messiah.

> For this purpose was the Son of God manifested, that
> he might destroy the works of the devil. (I John 3:8)

We have seen in the bribery attempt on Jesus that Satan claimed to have dominion over the great empires of the ancient world. In this verse, John unequivocally states that Messiah would destroy those works. Is there anyone so locked into a scientific or religious paradigm as to be incapable of recognizing the fulfillment of this extraordinary prediction?

Paradigm Lock

Academia's denigration of the ancient texts as myth has left most educators with very little knowledge of the historical figure called Satan. When I confront them with the correlation between the biblical prophecies of his demise and the collapse of that world, they attribute it to coincidence and ridicule my assertion that the Bible's "mythical Satan" played a role. Their preconceived beliefs blind them to both the diabolical and supernatural nature of these ruins so readily apparent to even a casual observer.

For instance, I will never forget my first trip to the Mayan ruins at Chichen Itza. The predominant building in this complex is a huge edifice called *the Castle*. This imposing stepped pyramid is dedicated to the Feathered Serpent, Kukulcan. Like the Egyptians, the Mayans designed it to reflect the dynamics of the solar system.[55] Archeologist Luis Arachi discovered that Mayan astronomers used the Castle as a celestial calendar. On March 21st and September 21st an incredible phenomenon occurs. At the precise time of the equinox, the design of the pyramid causes a series of triangular shadows to be cast on the side of the main staircase. As the sun follows its course, these shadows simulate the body of an undulating snake that connects with the serpent head built into the base of the staircase. In March the serpent appears to be descending, and in September it appears to ascend the pyramid. People come from around the world to view this phenomenon, which lasts for approximately three hours and twenty minutes.

There is a temple at the top of the Castle which can only be reached by climbing an almost vertical set of very narrow stone steps. My twelve-year-old son, Kevin, was with me on the trip, but he was so intimidated by the size of the pyramid that he was declining my invitation to make the climb. Only with my strong assurance for his

Author standing in front of the Castle in Chichen Itza, Mexico. Please note open mouths of serpents that adorn the foot of the staircase.

safety did he finally agree to accompany me. On the way up, I had to stop several times to reassure him because the steps are very narrow and there are no handrails to alleviate the anxiety generated by the imposing nature of the pyramid's construction. When we finally reached the top, Kevin would only sit on the floor, terrified as to whether or not he would be able to safely negotiate the descent. I too was unnerved, but I tried to conceal my discomfort so as not to exacerbate the situation. When Kevin had settled enough to allow me to explore the temple, I was again moved by the malevolent nature of these buildings. The temple itself is rather small when compared to the structure on which it sits. It has no windows, so it stays damp and cold even in the hot Mexican climate. Uncomfortable in this claustrophobic atmosphere, I made a quick study of the temple's construction and emerged from its gloomy darkness into the bright Mexican sun to rejoin Kevin. As we sat there trying to muster the courage to make the descent, I couldn't help but wonder what it must have been like for our ancestors who

The Pelota ballcourt just beyond the ceremonial green. The large building with three entrances attached to the ballcourt is the Temple of the Jaguars.

were summoned to the top of this structure to meet their fate at the hands of the depraved beings who ruled that world. The thought of being called to judgment before a god described as a feathered serpent was truly frightening. It is recorded that as many as ten thousand humans were sacrificed to these deities in a single ceremony and that they would cut out the hearts of their victims, consuming them while they were still beating.

As I looked at the view from the top of the pyramid, I could see in the distance the huge Pelota ballcourt. It is reported that a fierce game of sport was played on this court, where the losing teams would have their heads cut off by the winners. The large ceremonial green in the foreground is bordered on the right by the platform of the Skulls,[56] decorated on all sides with depictions of human skulls, and the platform of the Tigers and Eagles, adorned with depictions of vicious-looking tigers and eagles clutching human hearts in their claws.[57]

An examination of the ballcourt reveals the macabre nature of the Mayans as well as their advanced engineering skills. Standing at

The Temple of the Bearded man at one end of The Ballcourt.

one end of the arena is a large, corbeled ceremonial platform called *the Tribune*. 150 yards away, on the opposite end, stands the Temple of the Bearded Man. The Mayans designed the acoustics of the arena such that anyone speaking in a normal voice from the temple at one end can be heard distinctly on the platform at the opposite end. The field itself is 272 feet long and 199 feet wide. The two 27-foot-high lateral walls dwarf the tourists in the picture above, who are examining the magnificent embossed artwork on the lower sections. Please note the ring near the top of the wall, approximately 23 feet above the ground. It is thought that the object of the game was to get a ball through one of two such rings. The winners would receive all the jewelry worn by the spectators, while the severed heads of the losers were placed on the platform of the Skulls. I was sickened by the thought of the fierce competition that must have been evoked by this depraved form of mortal combat.

The view from the Castle was stunning, but the atmosphere produced by the history and decor of the complex left one feeling

Mayan deity with snakes encircling his neck; Cop-n, Honduras.

very uneasy. These negative vibes, combined with the great height of the pyramid, kept Kevin glued to his seat. Eventually, I managed to convince him that he could make the descent if he refused to look down and took one step at a time. When we finally reached the bottom, the evil nature of this empire continued to assault our senses.

The majority of the buildings at Mayan sites are decorated with perverse artwork ranging from grotesque creatures to angry-looking gods with snakes dangling from their mouths. These decorations are more ghoulish than stately and seem better suited to a house of horrors than to the major cities of a great empire. Art is a mirror of the soul, and these images provide us with a window into the true nature of the beings who ruled the ancient world. We don't find flowers or butterflies depicted in their cultures; it is always human skulls or sinister creatures like snakes, scorpions, spiders and dragons. The entrance to one building at Chichen Itza depicts a human in the jaws of a giant serpent. Suffice it to say that wholesome beings don't decorate their governmental buildings with hideous

Close-up of deity shows snakes emerging from his mouth.

and menacing artwork. Clearly, these places were not designed to make humans feel at ease.

The question that always plagues me when I research the ancient ruins is how the archeologists can fail to sense the overtly evil nature of these empires. The answer lies in their inability to communicate with their adversaries in theology. Ignorant of the fact that Satan had already been destroyed, leaders of the early Christian Church feared these relics. They believed that anyone who read the

Skull rack from Mayan ruins; Cop-n, Honduras.

pagan writings or studied their artifacts would become possessed by evil spirits, so they had them destroyed. An equally destructive backlash against this religious paranoia caused scientists to make the denial of Satan's existence a tenet of their atheistic cult. That belief has forced them to ignore the depraved nature of these civilizations clearly manifest in their own research.

Academia fails to recognize the evil nature of the ancients because they are incapable of perceiving any reality that contradicts

This intimidating staircase of 150 steps rises 125 feet to a temple at the top of the Pyramid of the Soothsayer; Uxmal, Mexico.

their atheistic beliefs. I call this syndrome a paradigm lock. The Darwinian paradigm dictates that scientists see the ancient empires as the work of primitive humans; therefore, the unmitigated evil of the ruins simply will not compute. However, anyone who is not required to view these relics through an atheistic prism will easily discern the difference between primitive culture and sophisticated evil.

In the next chapter we will see how academia's desatanizing of the ancients slowly evolved into a backhanded endorsement of the cultural side of these civilizations. This quasi-affirmation allowed modern leaders like Napoleon, Hitler and Mussolini to justify emulating these cruel dictatorships as an expedient in their efforts to restore the glorious empires of their ancient "ancestors." The disastrous results of that folly are currently being repeated in numerous countries around the globe. Instead of lionizing the demonic despots who ruled that world, educators should be honoring the one who liberated us from their tyrannical reign. Unfortunately, Jesus, like Satan, is anathema in the cult of academia. It is incredible that these two pivotal figures, around which the entire span of human existence revolves, have been summarily banned from consideration in the "scientific" quest for truth. Can you imagine the outcry from these atheists if theologians were to insist that their students totally ignore pagan history and deny the existence of the Pharaohs and Caesars? It is time for academia to restore both Satan and Jesus to their proper place in history and return the Bible to its proper place in academia. Short of that, we will never be able to solve the great mysteries of either the ancient or modern worlds.

THE NEW PARADIGM

There is extraordinary significance in the fact that the advanced civilizations of the Incas, Mayans, Egyptians, Romans and Greeks now lie in ruins. These powerful empires were part of the bribe that Satan offered Jesus on the mountain. If Messiah had accepted his proposition, those civilizations would be vibrant centers of knowledge, wealth and power, and humans would still be enslaved to the satanic forces that ruled the ancient world. Instead, Satan is dead and we humans are struggling to preserve the ruins of those cities as evidence that such civilizations once existed. In light of the overwhelming physical and sociological evidence confirming the fulfillment of the end of the world prophecies, we should formulate a new paradigm based on the biblical declarations that the Messiah of Israel would destroy Satan and the meek would inherit the earth.

As Schliemann's discoveries revealed that Greek "mythology" was real history, the collapse of the ancient world in the precise manner predicted in biblical prophecy proves that the Bible is an equally accurate historical account of God's dealings with men in the ancient world. Yet the erroneous beliefs of science and religion have so discredited biblical history that most people ignore this vital resource when they attempt to solve the mysteries of the ancient world. Authors like Tompkins and Daniken astonish us with the glorious relics and sophisticated intelligence of the pagan civilizations, but they are forced to admit that they have no solid explanation for their demise. Without knowledge of the biblical pieces of the puzzle—the second coming of Messiah, the Rapture of the disciples, the fall of Jerusalem and the death of Satan—this mystery will never be solved. The time has come to restore the Bible to its proper place as a closed history of God's interaction with men in the ancient world. That done, we can put aside the confused and divisive doctrines of science and religion, and embrace the uplifting new paradigm implicit in the fall of the gods and subsequent rise of man to rule the earth.

Did the world end two thousand years ago? The answer is yes! The Almighty God Himself caused the fall of Jerusalem and the demise of the ancient world, and no man, or group of men, no matter how sincere, will ever be able to put that Humpty Dumpty back together again.

WHAT'S NEXT?

Many people ask me, "If the second coming of Messiah has already occurred, then why are we still here? Did God just abandon us?" The answer is no. God did not abandon man, He simply solved our immediate problems and prepared for our future. Having accomplished all He planned to do on earth, God did end His communication with us, but in the Scriptures, He told us what to expect in the future.

> For as in Adam all die, even so in Christ shall all be
> made alive. (I Cor. 15:22)

The God of Abraham has great love for mankind, and, as this scripture declares, He has already solved the problem of death brought on all men by the transgression of Adam. According to this verse, all men will be given a new life through the work of the Messiah. What will that life be like? We get a glimpse of it in the following verses from the writings of Isaiah in the Old Testament and the apostle Paul in the New Testament:

> For since the beginning of the world men have not
> heard, nor perceived by the ear, neither hath the eye
> seen, O God, beside Thee, what He hath prepared for
> him that waiteth for Him. (Is. 64:4)

> Eye hath not seen, nor ear heard, neither have entered
> into the heart of man, the things which God hath
> prepared for them that love Him. (I Cor. 2:9)

In celebration of the two-thousand-year anniversary of the arrival of the Messiah, let us hold a symbolic funeral for Satan and celebrate our liberation from his tyrannical demons, who terrorized our ancestors in the ancient world. Let us also pledge to cut the philosophical umbilical cords that keep modern man tethered to the anachronistic rituals and jihads of the ancient world. As we enter the 21st century, we must come together and, as promised by King David in Psalm 37, begin to delight ourselves in the peace and freedom afforded us by the fall of the gods.

In light of all the complex hypotheses put forth as explanations for the demise of the ancient world (visiting alien forces returned to their planet, disastrous climactic change, disease, immorality), I am sure that many in academia will decry the simplicity of my solution. In response, I invoke the law of Occam's Razor. Named for the 14th-century British philosopher who first proposed it, this widely accepted law of science holds that the simplest explanation that fits all of the facts is probably the correct one. Thus, the simple truth.

III. Humans Inherit The Earth

The historians Josephus and Tacitus recorded eyewitness accounts confirming that, precisely as predicted by the biblical and pagan prophets, a great war of the gods took place in A.D. 70. That battle severed our ability to communicate with those supernatural beings, resulting in the collapse of the advanced civilizations of the ancient world and the subsequent rise of man to rule the earth. Our failure to properly comprehend the implications of that great cataclysm plagues us to this day. The ramifications of that mistake are readily apparent in the Holocaust, communism vs. capitalism and the East-West struggle, the bombing of the World Trade Center, the ashes and tombstones of Bosnia, Belfast and Beirut, World Wars I and II, the Korean War, the Vietnam War, the Six Day War, the Gulf War, the downing of Pan Am flight 103, the terrorist bombings in Israel, the ideological machinations of the political Left vs. Right and a million other insanities, but at the heart of it all is the mindless ideological jihad between science and religion. In this chapter I will present a chronology of how our ignorance of the truth inherent in the collapse of the ancient world has tortured us all through history, permeating every aspect of modern life.

As we have seen, prior to the fall of Jerusalem, powerful beings ruled the earth. These gods and their demigod offspring (the pharaohs, emperors, royal incas and kings) controlled all aspects of life in the ancient world. The demigods were produced by a union between a human virgin and an ancient god. That demigod offspring would become the earthly ruler of an ancient civilization, and at his death, his firstborn son (through the system of primogeniture) would receive the mantle of his father. All the other offspring of the ruler were part of the aristocracy, but they did not receive the power that was transmitted to the firstborn son. The great knowledge and physical stature of these half-human demigods gave them total dominion over ordinary men, whom they kept in a position of servitude.[1] Some humans of special aptitude were chosen to work in the palaces and temples of these god-men, but, for the most part, the role of humans in that age was beneath the role that animals play for mankind today. The demigods bred them like cattle and used them as slaves. They could be thrown to lions for sport, forced into mortal combat in amphitheaters, worked to death on meager rations or drafted to be slaughtered in the endless wars of the ancients. When the gods were destroyed by the second coming of Messiah, the earthly demigods lost communication with the source of knowledge and power that allowed them to maintain their positions. As the last generation of these god-men died off, their human descendants in the aristocracy usurped their positions and, through pomp and ceremony, deceived the rest of mankind into believing that they were the rightful heirs to these ancient thrones. Eventually, this hoax was exposed, and many of these pretenders were forced either to abdicate, as can be seen in the case of the Emperor in China, the Czar in Russia and the King in France, or to assume powerless positions such as those of the king or queen in England and the emperor in Japan.[2] Many of the remnants of these ancient theocracies, like the papal system in Italy, still operate in the guise of religion, but they too are slowly being forced to admit that they are unable to communicate with the gods.

When mankind was bequeathed the role of ruler of the earth by the demise of the gods, he was lost. With the exception of the few educated slaves who had served in the palaces and temples of the ancients, the god-men of that world had not shared their wisdom with man. Consequently, the great knowledge of astronomy, geometry, mathematics, medicine, physics etc. used by the ancients was all but unknown to the humans of that age. A good example of the suppression of man's knowledge can be seen in the case of the Mayans in what is now Mexico and much of Central America. The Mayan people had no knowledge of either the shape of the earth or the workings of the solar system; they were trained by their rulers to plant and reap when the sun reached certain steps on their temple. Yet we now know that the Mayan leaders had an accurate calendar thousands of years before the modern era, indicating that they had full knowledge of the solar system. When the educated descendants of these demigods died off, much of this knowledge was lost. Later generations of Mayans were incapable of reading the hieroglyphics of their ancestors, nor could they comprehend the intricate calendar that the ancients had left behind. The great wealth and monumental palaces and temples of these demigods had little meaning for them either, so they just walked away, preferring the simple life of the jungle to the ostentatious splendor of these ancient cities. Because the economic system of the ancients did not require their subjects to interact with neighbors, many of their cities were located in remote parts of the planet, far from the oceans that were necessary for transportation. This made them environmentally and economically uninhabitable for humans in the modern world.[3] As a result, many were abandoned to the elements and were soon overgrown by dense jungle or buried in sand by desert storms.

With the fall of Jerusalem, the prophecies of David and Jesus that the meek would inherit the earth were fulfilled, and the gods and demigods of the ancient theocracies no longer controlled human destiny. Now, the system of primogeniture, which had provided for the orderly ascension of the firstborn son of a demigod ruler to the royal throne, was ended. All access to the supernatural powers that ruled the ancient world had been severed by the second coming of Messiah, leaving the future firstborn sons of these sovereigns fully human. Nevertheless, old habits are difficult to break. More than 1,900 years have passed since the end of that world, but in many countries (like Japan and England) people continue to believe that their leaders are ordained by divine providence. There are several reasons for this phenomenon; chief among them is the fact that the members of the aristocracies in these civilizations were loath to relinquish their lucrative positions. In an effort to retain their power, the aristocrats and priests continued to elevate the firstborn sons of the royal family to the throne, regardless of their abilities. This became ludicrous when the premature death of a ruler forced them to place a small child on the throne of a vast empire under the pretense that the gods had ordained his ascension. For example, the last emperor of China was just three years old when he took the throne in 1920.

The historical accounts of the turmoil in Rome after the suicide of Emperor Nero in A.D. 68 provide critical insight into the rise of the first human to rule the earth. Because he was childless and had not chosen a successor, Nero's death severed the royal bloodline and left the empire without an heir apparent. This plunged Rome into a terrible civil war, which came at a particularly portentous time because Nero had recently sent the great Roman General Vespasian to end the civil war in Judea. Vespasian and his son Titus were, at that very time, preparing for the assault on Jerusalem. Now Rome itself was in great disorder, and a series of revolts quickly led to the overthrow of three emperors, Galba, Otho and Vitellius. At this point, Vespasian's commanders began urging him to take the empire, but because he was of low birth (and being a truly humble man), he refused, determined to await orders to renew the war from a rightful emperor. Finally, when his officers threatened to revolt if he didn't take the throne, he acquiesced and declared his willingness to try to restore order. Vespasian was proclaimed emperor by his troops on Jewish soil and quickly received the support of other prominent generals. He then returned to Rome, ended the civil war and took control of the empire.

The Roman Senate supported him as emperor though he had no connection to the Patrician aristocracy, making him the first fully human being to rule on earth. Correlating Vespasian's rise to emperor with the cause of the civil war in Jerusalem, Josephus writes:

> But now, what did most elevate them [the seditious Jews] in undertaking this war was an ambiguous oracle that was also found in their sacred writings, how, "about that time, one from their country should become governor of the habitable earth." The Jews took this prediction to belong to themselves in particular; and many of the wise men were thereby deceived in their determination. Now, this oracle certainly denoted the government of Vespasian, who was appointed emperor in Judea.
> (*Wars of the Jews*, Bk. 6, 5:4)

This is an incredible statement because Josephus, who was a high priest, declares that Vespasian's rise to power was the fulfillment of the Jewish prophecy that foretells the rise of an earthly ruler. Modern theologians declare that this prophecy is yet to be fulfilled. Only one of these positions is correct. Let us look at some of the evidence supporting Josephus' claim. There are many incredible facts surrounding the rise of Vespasian. He was not a member of a Patrician family, from which, according to Roman law, an interrex[4] or a new emperor had to be chosen.[5] Galba, Otho and Vitellius were all Patricians and, with senate approval, had sought to make themselves emperor, but none of them was able to maintain the throne after Nero. In contrast, Vespasian refuses to elevate himself, even at the behest of his troops. Finally, frustrated by his humility, they threaten to kill him if he doesn't take the diadem. When he acquiesces to their demands, he is named emperor not in Rome, as had always been the case previously, but on Jewish soil, as prophesied.[6] At that point, all the other military leaders endorse him, and the Senate is forced to go along. This was an unprecedented turn of events. In addition, shortly before

Vespasian returns to Rome as emperor in the summer of A.D. 70, peace is restored,[7] and, under what became known as the Flavian dynasty, it is maintained for a long period of Roman history.[8] I am in complete agreement with Josephus: Vespasian was the prophesied first fully human being to rule on Earth.

In the ancient world, the fact that a ruler was ordained by the gods was a critical part of his ability to intimidate his enemies and win the confidence of his subjects. The people knew that if a leader died prematurely, it meant that he had fallen out of favor with the gods. This was a terrible omen for the entire population. If neighboring countries discovered that the gods had forsaken your cause, they would become emboldened and attack. Not having a divinely ordained leader would leave the populace feeling vulnerable and might even encourage insurrection or civil war. To avoid this problem, the supporters of these human pretenders to royalty would often arrange for miraculous signs to accompany the coronation of a new leader.

Vespasian was willing to govern the empire but felt uncomfortable when, in order to reassure the populace, the aristocracy wanted to announce that he possessed divine powers. Tacitus relates a particular incident that is very revealing.

> In the course of the months which Vespasian spent at Alexandria, waiting for the regular summer winds when the sea could be relied upon, many miracles occurred. These seemed to be indications that Vespasian enjoyed heaven's blessing and that the gods showed a certain leaning towards him. Among the lower classes at Alexandria was a blind man whom everybody knew as such. One day this fellow threw himself at Vespasian's feet, imploring him with groans to heal his blindness. He had been told to make this request by Serapis, the favorite god of a nation much addicted to strange beliefs. He asked that it might please the emperor to anoint his cheeks and eyeballs with the water of his mouth. A second petitioner, who suffered from a withered hand, pleaded his case too,

also on the advice of Serapis: would Caesar tread upon him with the imperial foot? At first Vespasian laughed at them and refused. When the two insisted, he hesitated. At one moment he was alarmed by the thought that he would be accused of vanity if he failed. At the next, the urgent appeals of the two victims and the flatteries of his entourage made him sanguine of success. Finally he asked the doctors for an opinion whether blindness and atrophy of this sort were curable by human means. The doctors were eloquent on the various possibilities. The blind man's vision was not completely destroyed, and if certain impediments were removed his sight would return. The other victim's limb had been dislocated, but could be put right by correct treatment. Perhaps this was the will of the gods, they added; perhaps the emperor had been chosen to perform a miracle. Anyhow, if a cure were effected, the credit would go to the ruler; if it failed, the poor wretches would have to bear the ridicule. So Vespasian felt that destiny gave him the key to every door and that nothing now defied belief. With a smiling expression and surrounded by an expectant crowd of bystanders, he did what was asked. Instantly the cripple recovered the use of his hand and the light of day dawned again upon his blind companion. Both these incidents are still vouched for by eye-witnesses, though there is now nothing to be gained by lying. (*The Histories* 4 "Signs And Wonders")

There are many flaws in this scenario. Even Vespasian finds the suggestions that he possessed godly powers to be humorous. Can it be that a man could possess the divine power to heal and not know it? If the claims of the invalids were true and Serapis did tell them to go to Vespasian for a miracle cure, we are left with an obvious question: Why didn't Serapis inform Vespasian that he possessed the power to heal? Notice also how the royal entourage is encouraging Vespasian with flatteries. I think that this was a setup designed to reassure the people that the gods were supporting their new emperor, even though he was not of royal blood. Tacitus' last statement about lying indicates that there were some who doubted the veracity of this incident. Though all the emperors from Vespasian to Aurelius[9] (save Domitian) were deified after death, unlike the ancient Caesars, they were known to deny immortality and made light of the concept that they were gods.[10] Over time, most of these pretenders to royalty have either been exposed or overthrown by revolution, as man is slowly coming to terms with the fact that humans are the supreme beings of the modern world. For the most part, we have replaced the ancient monarchies and theocracies with democratic governments, leaving religion the one place where this pretense of ancient power continues unchallenged.

THE DARK AGES OF THE NEW WORLD ORDER

With the destruction of Jerusalem, the world was plunged into a state of chaos. All communication with the gods was severed, and as the great men of that age slowly died off, most of the advanced science and academic prowess of the ancients was lost. This was the beginning of the dark ages. Man was now alone on the earth. Not unlike the student monitors in an elementary school class where the teacher has left the room, the new human rulers were on shaky ground. As pretenders to royalty, they lacked the supernatural power that their predecessors had used to intimidate their subjects and were dependent on a supporting cast to maintain their illusion of deity. Men lived in great fear, unable to comprehend their environment or the advanced mathematics, astronomy and general science detailed in the writings of the ancient scribes. This ignorance caused men to begin to speculate about the size and shape of the earth, as well as where it was located in the universe. These fledgling attempts to understand our environment were the beginning of modern (or human) science and were largely the work of scholars in the newly developing Judeo-Christian religions. The rise of modern religion facilitated the first great shift of power from the monarchies to the Church. This was the beginning of modern politics. Eventually the Church became more powerful than the state, causing a secular backlash and the rise of our current systems of government.

THE RISE OF MODERN RELIGION

After the fall of Jerusalem, the Jews (like the pagan theocracies) refused to face the magnitude of the disaster that had befallen them. The Sanhedrin[11] and other members of the aristocracy knew that Jerusalem was the city where God had told David to place His tabernacle (dwelling) on earth. Its total destruction, without the coming of a prophet like Jeremiah to reassure them that they would return to the land, was an ominous sign indicating that God had removed His hand of protection from the Jews. Nevertheless, they simply could not face the fact that God Himself had ended their world in the precise manner described in their own prophecies. It was, understandably, a truth just too difficult to face. In the Diaspora, the Jews tried to pretend that nothing had changed. Religious leaders developed a system of scriptural interpretation called the Talmud, and by this means they kept the Scriptures away from the masses, thus avoiding the difficult questions concerning the prophesied coming of Messiah and the end of the world.

Unfortunately, it was no longer possible for the Jews to continue as though nothing had changed. The destruction of Jerusalem in A.D. 70 was unlike any previous event in their history. During the Babylonian captivity, God told King Nebuchadnezzar to allow the Jews to keep their lineages and live a separated life under their own law so that they could eventually return to Jerusalem as a distinct people. When God allowed the Romans to conquer Jerusalem the first time, they again remained a separated people, and Rome permitted them to judge themselves under Jewish law. Their lineage and laws were the essence of ancient Israel; now the Romans had burned those records, and the Jews who had survived the war were already divided into so many splinter groups that they couldn't agree on who was, and who was not, a Jew. No longer able to judge themselves by their own law, they would now become subject to the laws of the kingdoms to which they were dispersed.

While modern Jews may well be the descendants of the ancient Israelites, those lineages cannot be authenticated. And even if they could be authenticated, they would have no legal standing here in the modern world. The descendants of the ancient Vikings, Incas, Egyptians, Romans, Druids, Mayans and Greeks do not try to claim land rights or privilege based on ancient law. They have all accepted the reality implicit in the demise of that world. For the sake of peace, it is time for the Jews to do the same.

The Jewish lineage problem was compounded by the rise of a new sect of Jews called Christians. These were the insincere Jews that had followed Messiah Jesus for worldly reasons. He referred to them as false apostles and prophesied that they would be left behind when he supernaturally removed his faithful followers from the earth.[12] Like the traditional Jews, these Christian-Jews were unable to face the obvious implications of the destruction of Jerusalem. They, too, decided to pretend that God was still dealing with Israel. However, unlike the traditional Jews, the Christian-Jews were proselytizers. They took their message beyond the lineage of Israel to the heathens and quickly developed a large following. This created a bitter ideological feud, resulting in bloodshed. These conflicting denominations of Judaism continued to quarrel over petty doctrinal disputes, eventually dividing into the diverse denominations we see today: Orthodox, Conservative, Reform, Catholic, Lutheran, Born Again, Evangelical, Episcopal, Shiite, Sunni, Jim Jones' Peoples' Temple and Branch Davidian, to name just a few. Though all of these religious sects have their roots in the Jewish Scriptures, after A.D. 70 they were merely pretenders to the place of God's chosen people: the ancient Israelites. The violence associated with that ideological charade plagues modern civilization to this very day.

With the destruction of Jerusalem, Rome became the center of Jewish civilization. In A.D. 130 the Roman Emperor Hadrian (117–138) tried to end Judaism once and for all. In an attempt to assimilate the Jews, he issued an edict prohibiting circumcision, the reading of the Jewish law and observance of the Sabbath. He also gave orders to rebuild Jerusalem as a Roman colony.[13] The effect of Hadrian's purge was to goad the Jews into a last, desperate uprising. In the year 132, under Simon Barcocheba, who took the title "Prince of Israel," the revolt began. Barcocheba was hailed as Messiah by the famous Rabbi Akiba, inciting the Jews to support the rebellion. With more than 200,000 insurgents, he liberated Jerusalem and other Roman strongholds in Palestine. Writing of the revolt in *The Origins of Christianity*, Archibald Robertson says:

> But Trajan's[14] generals twenty years before had done their work too well for the new revolt to be anything but a forlorn hope. An immense army under Hadrian's best general, Severus, was massed against the rebels. Even so, not until 135 was Jerusalem recaptured, Simon killed, the last rebel stronghold at Bether, south-west of Jerusalem, taken and the war, at enormous cost in life on both sides, brought to an end. The conquerors massacred men, women and children in thousands, sold thousands more dirt-cheap in the slave markets, and flayed Akiba alive. Jerusalem was rebuilt as a Roman city under the name of Aelia Capitolina; a temple of Jupiter Capitolinus was erected where the Jewish temple had stood; and Jews were forbidden to come within sight of the city on pain of death. One result of this revolt was the final separation of Christianity from Judaism. Even Jewish Christians had refused to back Barcocheba; and the gentile churches had been too well dosed with anti-Jewish propaganda by their leaders to run into danger. There was now no longer room, as some first-century

Paulinists had thought, for a Petrine "gospel of the circumcision" beside the Pauline "gospel of the uncircumcision." The Jews, three times in seventy years crushed by the military might of Rome, were in Christian eyes plainly under a curse. Jewish Christianity became a heresy. (Robertson, p.183–4)

This war finalized the split between the traditional and Christian-Jews, leading to an acceleration in Christian proselytizing. As a result, Christianity experienced a rapid growth in converts from the ranks of the pagans of Rome. Feeling threatened, the pagan sects instigated a government crackdown, and the growing Christian Church was forced underground. This persecution became barbarous, and many Christians, even those with Roman citizenship, were thrown to lions for sport or burned alive on crosses. The violent repression continued unabated until the reign of Emperor Constantine (288–337).

In 300 Constantine claimed to have seen a vision in the clouds accompanied by a voice which told him that by the sign of the cross he would conquer. As a result, he made Christianity the official religion of the empire. Now the tables were turned. The Christian Church enjoyed the blessing of the emperor, while the pagans and traditional Jews faced persecution. To say that the Christians were less than magnanimous in victory would be an understatement. The pagan religions were declared satanic, and traditional Jews were condemned as Christ-killers. In order to ease the transition from paganism to Christianity, many of the rituals, holidays and customs of the pagans were incorporated into the still-developing Christian religion. The Winter Solstice became Christmas; the pagan spring festival, named for the Babylonian moongoddess, Ishtar, became Easter, and the pagan statues of Ishtar and her baby, Tammuz, were reinscribed to read Mary and Jesus. With the pagans being forced to convert and non-proselytizing, traditional Jews limited to cloning new members, Christianity experienced explosive growth, while the Jews and pagans quickly became persecuted minorities.

After the death of Constantine, his sons ruled for a time. Then, in 361, the army proclaimed Julian emperor. Julian was called "the Apostate" because he tried to uproot Christianity and restore paganism. His efforts failed and only served to strengthen the Church. Now the conglomeration of Christianity, Judaism and paganism took the title Catholic, or universal, Church. This new religion amassed great wealth and became a major source of money for science and the arts. Emperors, kings and rulers of every stripe came and went, but the Church remained. Christianity expanded virtually unchallenged until the rise of yet another religion based on the Jewish Scriptures: Islam.

In the seventh century, Mohammed claimed to have been called of God to build a new chosen people based on the lineage of Ishmael, the half-brother of Abraham's son Isaac. This new religion flourished in the Arabic civilizations and soon became a force to be reckoned with. Several wars between Islam and Christianity, called Crusades, were fought for control of the Holy Land—the area once occupied by the ancient city of Jerusalem. By the time these wars were over, uncounted thousands of believers had given their lives in the name of either God, Allah, Mohammed or Jesus Christ. If these conflicts can be said to have proven anything, it was that the God of Abraham wasn't supporting either side.

By the 11th century, Christianity had spread over most of Europe. It was an involuntary association: a person was born into it and dared not leave. It taxed and judged him, and it cared for him in sickness and poverty. Originally just a spiritual organization, the Church was now becoming a super-state that could enforce its will on the governments of Christianized countries. In 1198 Pope Innocent III vigorously asserted that the pope was superior to all kings and emperors, and many rulers, fearing his political influence, were forced to admit his claim. Eventually, all the rulers of Christian countries had to present themselves at the Vatican for a papal blessing before they could rule. This is dramatic evidence of the loss of royal authority due to the collapse of the ancient world. Under the original system, the royal family itself possessed the ability to communicate with the gods and perform miracles. That is why Vespasian's entourage tried to reassure the people by having him exhibit the ability to tap into the power of the gods. The fact that the Church had now usurped that role indicates the weakness of the post-A.D. 70 royalty. Had these modern rulers possessed the powers of the ancient pharaohs, emperors, royal incas and kings, the Church could never have taken such liberties. This change gave the Church great power.

ANTI-SEMITISM

By the end of the first millennium, the growing power of the Church had all but eliminated the pagan religions on the European continent, and the Jews became the main focus of Christian persecution. By refusing to accept Jesus as the Messiah, they were, in effect, contradicting Christianity. Considering the size of the Church at that time, theirs was not an enviable position. Some Jews, seeing the handwriting on the wall, did convert to avoid physical and financial ruin; however, most were unwilling to forgo their beliefs and subject themselves to the humiliation of baptism. Making matters worse was the fact that Judaism is based on a chosen lineage; therefore, Jews were forbidden to intermarry with non-Jews. This intolerance was perceived as an insult that only compounded the injury of their unwillingness to be baptized. In retaliation, Church leaders took a very aggressive anti-Semitic stand. They declared the first-century Jews guilty of having killed the son of God, and they indoctrinated Christians with the belief that all Jews who refused to convert would suffer eternal damnation merely for being the descendants of those who had committed this terrible deed. In turn, they made it a grievous sin for a Christian to marry a Jew, leaving the defector subject to the same fate that awaited all of these "Christ-killers." Eventually, the Jewish religion itself was declared to be satanic, and the Church hierarchy even went so far as to claim that the Jews had horns.[15] In fact, a Michelangelo statue of Moses,

Monument of Pope Julius II: Moses

commissioned by the Church and still on display near Rome, depicts him with two horns protruding from his head.

The idea that Jews were guilty of having killed the son of God made those that refused to convert seem all the more sinister, creating intense anti-Semitism among Christians. This led to the entire population of Jews being expelled from several Christianized countries. Most notable of these were the mass expulsions from England in 1290 and from Spain in 1492. This perverse doctrine ignored the fact that Jesus himself had declared that no man could take his life.

> As the Father knoweth me, even so know I the Father:
> and I lay down my life for the sheep.
> And other sheep[16] I have, which are not of this fold:
> them also I must bring, and they shall hear my voice;
> and there shall be one fold and one shepherd.
> Therefore doth my Father love me, because I lay down
> my life, that I might take it again.

No man taketh it from me, but I lay it down of myself.
I have power to lay it down, and I have power to take
it again. This commandment have I received of my
Father. (John 10:15–18)

Though a case can be made that the leaders in Jerusalem and their followers forced Pilot to crucify Jesus, this was a small percentage of the population.

And they [the Jews] were instant with loud voices,
requiring that he might be crucified. And the voices of
them and of the chief priests prevailed.
And Pilate gave sentence that it should be as they
required. (Luke 23:23–24)

In this quote it is clear that the Romans were as guilty as the chief priests because Pilot used Roman soldiers to execute a man he knew to be innocent. Notwithstanding, Jesus asked God to forgive them all because what they did was done in ignorance.

Then said Jesus, Father, forgive them; for they know
not what they do. (Luke 23:34)

This quote declares that the Jews and Romans involved in this incident were simply pawns being manipulated by forces far more powerful than mere humans. The crucifixion of Jesus was part of a much larger struggle between supernatural entities that were vying for control over humanity and the earth.

THE GREAT SCHISMS

The lack of real power in Christianity caused schisms which began to split the Church itself. In 866 the Greeks were the first to separate from Rome, forming the Greek Orthodox Church. This was the beginning of doctrinal fracturing within the Church, and it has continued all through history. Intellectual ferment between the 12th and 15th centuries produced additional internal rebellions against Church doctrine. These heretics became so numerous that the Church had to organize crusades against them in southern France. In 1520 Martin Luther was excommunicated by the Vatican, resulting in the creation of another major schism: the Lutheran Church in Germany. The fracturing of the Church continued, and in 1534, having been refused an annulment of his marriage to Catherine of Aragon, King Henry VIII formed the Church of England as a separate entity from Rome. At about this time still another schism, based on the teachings of John Calvin, arose in Switzerland and Scotland. To counter this seditious behavior, the Church instituted the Inquisition as a special court to prosecute these "dissidents, witches, demons and heretics."

EXPLORATION AND PROPAGATION OF THE FAITH

In the 15th and 16th centuries, Spain controlled a great Christian empire, and many explorers like Cortés, Pizarro and Columbus were supported by the Church and the Spanish Crown. It was during this period that Spain conquered the Aztecs, Incas and American Indians, forcing them to convert to Christianity. The Church (being totally ignorant of the fact that Satan had been destroyed in the great battle of the gods that accompanied the collapse of the ancient world) considered the pagan religions to be satanic, so the monks accompanying these expeditions burned any literature they found and forcibly ended the practice of the ancient rituals. The destruction of those writings and traditions was a devastating loss to future research on the ancient civilizations of North and South America and the powerful forces who ruled that world.

THE RENAISSANCE

Between the 14th and 16th centuries, mankind was totally confused. Holy wars were raging all over the world, and tens of thousands of people were being tortured and killed in the name of God. In the New World, entire civilizations had been conquered and forced to convert from paganism to Christianity. This brutality, combined with the repression of philosophical freedom on the European continent, fostered an atheistic secular revolution which is commonly referred to as "the Renaissance." Intellectuals sought to revive the classics of antiquity as a means of finding an alternative to the dogmatic teachings of the Church. They became known as humanists because classical Greek and Latin literature was referred to as the humanities, or the more human literature in relation to the sacred literature of the Bible and Church dogma. This presented a problem for these atheists because many ancient writers are unequivocal in ascribing their work to godly inspiration. For example, both Homer and Virgil plainly state that the voice in their writings is not their own, but that of a deity. These writers are considered to be the pinnacle of literary achievement, and academicians still ponder how they evinced such great wisdom and literary skill so early in history. Conceding that modern writers can't hold a candle to works like *Iliad* and *Odyssey*, they voice confusion as to how these men did not appear to need to hone their craft, but achieved perfection at the outset. Meanwhile, the humanists simply ignored the explanation for this phenomenon given by the writers themselves because it contradicted their atheistic beliefs. To avoid the obvious flaw in their position, they declared the works to be myth and studied them as examples of great fiction.

As atheistic scientists created the ropes, wedges and rollers theory to avoid the obvious supernatural implications of the Great Pyramid, atheistic academicians were now forced to reduce the phenomenal accounts of Homer and Virgil to human fantasy. Unfortunately, ascribing this great literary skill to these writers forced them to come up with an explanation for the extraordinary absence of this intellectual prowess during the great ignorance of the dark ages. Their unwillingness to consider a supernatural explanation for the collapse of the ancient world led them to decide that the rise of authoritarian Christianity and its destruction of some pagan literature was responsible.[17] Blaming their age-old nemesis, the Church, solved two sticky problems for these atheists. On the one hand, it provided a human explanation for the great loss of ancient knowledge, and on the other, it allowed them to continue to ignore the supernatural aspects of the collapse of the ancient world. When it later became apparent that the Church could not possibly have been responsible for what proved to be a worldwide loss of knowledge and prowess, they changed their story and began to blame human and natural events like earthquakes, wars and famines. This steadfast denial of the supernatural only exacerbated their deteriorating relationship with the Church and kept them from discovering the true cause of the demise of the ancient world.

THE CULT OF SECULAR HUMANIST SCIENCE

The classical revival was a significant event in history. Prior to 1500, European life and culture was centered in the Church. Now, many intellectuals were beginning to look to the ancient writings, and European culture was becoming more secularized. Enthusiasm for the antiquities of Greece and Rome was carried over from literature to art and spread throughout the continent. A new secular humanist culture emphasizing man's carnal nature and his works was displacing the old Church culture that stressed the spirit and the world to come. If converts to this atheistic belief system were to fully sever the spiritual umbilical cords that tethered them to the old Church culture, secular humanism would have to be endorsed by an intellectual authority with ethical standards equivalent to those of the Church. The humanists found that affirmation in the cult of empirical science.

During the Renaissance, the Church-science relationship changed. Frustrated by the restrictions which forced them to bend their research to make it harmonize with the Church's theological interpretation of the Scriptures, secular scientists began to challenge the tenets of religion. This placed new emphasis on reason as a means of finding truth. The works of Aristotle became an important resource along with Arabic science, which had been discovered during the Crusades and from the Moors in Spain. Medical disasters caused by the introduction of gunpowder into warfare and rampant epidemics of the Plague forced a revival in medicine. Many of the ancient medical arts, banned by the Church as witchcraft, were rediscovered and put to new uses. In 1543 Andreas Vesalius, a Belgian physician, published *De Fabrica Corporis Humani*, in which he revealed (among many other things) that men did not have fewer ribs than women. This debunked the Church's teaching that men had lost a rib as a result of the creation of Eve from Adam in the Garden of Eden.[18] His book is considered a seminal work in the revival of ancient medical science. However, the most devastating blow to Church science during this period occurred when the Polish astronomer Nicholas Copernicus published *De Revolutionibus Orbium Caelestium*, his theory on the heliocentric solar system.

In the early 1600s, Galileo originated the experimental method of testing hypotheses which remains the cornerstone of modern science. His work disproved some of the postulations of Ptolemy and caused him to embrace the heliocentric solar system. In 1609 he developed a telescope that allowed him to study the heavens, and he was able to convince himself of the accuracy of the Copernican model. This led him to publish his *Dialogue Concerning the Two Great Systems of the World*, attracting the attention of his fellow Church scientists, who had embraced Ptolemy's geocentric universe but rejected Copernicus. These scientists had Galileo called before the Inquisition, where he was found guilty of heresy. They forced him to retract his theories and gave him an indefinite imprisonment. Though he was not kept in close confinement, his work was finished.

The persecution of Galileo was a turning point in history. It became the catalyst for a new science, totally separate from the Church. This was the secular "higher power" that would supply the affirmation so desperately needed by the humanists of the Renaissance. The god worshiped by this new cult would be human reason. Instead of being required to conform to Scripture and "moral rectitude," as was the case with Church science, all theories would be judged by their compliance with amoral facts. Before long, atheistic science was in the driver's seat, and truth and fact were no longer the private domain of the Church. Now these verities were claimed as the core principles of the new secular science, while Church scientists were relegated to interpreting the "mythical" Scriptures. The separation of science from the Church began a long, slow process of deprogramming mankind from the erroneous and dogmatic teachings of the Church scientists. Secular humanists saw themselves in the role of physicians delivering a curative medicine to a sick society. However, as with all remedies, at times the side effects can be more destructive than the illness they are designed to cure. The same superior attitude that led religious leaders to assume that they had the right, if not the duty, to rule in God's absence would now infect the scientists, producing equally disastrous results.

The new emphasis on Galileo's system of research produced discovery after discovery in rapid order: Mathematics, Physics, Medicine, Astronomy, Chemistry, Biology—all experienced great leaps of progress. Johannes Kepler's theory of elliptical orbits, combined with Isaac Newton's work on gravity, refined the Copernican solar system and brought an end to the geocentric universe. An important aspect of this rapid growth in human knowledge is that modern science takes credit for these "new discoveries," while, in fact, they are rediscoveries. As we have seen in the previous chapters, the ancient civilizations had all of this knowledge and much, much more several thousand years before the collapse of the ancient world left us in the deep ignorance of the dark ages. No doubt most of this information was readily available in the ancient writings, but it was lost due to the preconceived beliefs of religion in the early modern world. On the other hand, the ignorance of the atheistic scientists has been equally destructive. By declaring the texts to be myth, they denigrated the ancient writings, both biblical and pagan, nullifying their value as a reservoir of ancient knowledge. The Church has long since ceased burning books, but the atheists continue to censure these works, compounding the damage done by the clergy. This denigration of the historical texts by the competing cults of science and religion has not only kept us from using them to solve the mysteries of the ancient world, but those of the modern as well.

THE MODERN AGE

Berlin and Paris were the centers of the philosophical revolution. Great minds flocked to the universities of these cities, and several competing schools of philosophy began to take shape. The struggle between these secular thinkers and the Church was so intense that the doctrinal differences within the fledgling scientific community itself went somewhat unnoticed. However, these internal rivalries caused an equally destructive split between the radical atheistic philosophers of Paris and the more moderate agnostic[19] scholars in Berlin. Unfortunately, the atheists prevailed in this struggle and, over time, their philosophy engendered the ideological and political machinations that led to the Holocaust and an arms race that would bring the world to the brink of nuclear destruction.

Paris was a hotbed of revolutionary ideas where great thinkers like Voltaire, Hume and Rousseau were welcomed and celebrated. These were the Empiricists. They advocated an atheistic science rooted in the concept that all things were governed by a set of natural laws, which could only be discovered by examining empirical data. During this same period, the Germans forged an equally famous school of philosophy based on the teachings of Kant, Fichte, Schelling and Schlegel. Unlike Paris, which was still under the repressive yoke of the Catholic Church, Berlin had already experienced the reformation of Luther. This gave the Germans a small degree of intellectual freedom; consequently, they took a less reactionary view of biblical and pagan history. In contrast to the virulent atheism of Paris, German scholars advocated the more open-minded metaphysical historicism of Herder and Hegel. They proposed that while the application of empirical data might be valid in natural law—allowing us to understand why water freezes and melts at specific temperatures, for example—it was insufficient for the task of solving the more complex philosophical questions of human existence. Hegel taught that human intellect was a slowly evolving phenomenon that could only be understood in the context of the preceding eras. He spoke of a universal spirit that shaped history and moved men to ever higher levels of social consciousness. The Hegelanists believed in a non-violent evolution of knowledge and trusted that all men, even the clergy and monarchs, would accept the truth when properly educated to comprehend the need for change. This philosophy was dramatically different from that of the Empiricists in Paris, who had rejected the gods described by the biblical and pagan historians and were at that time preparing for the violent overthrow of the French monarchy.

Hegel challenged the atheists, accusing them of a dangerous fanaticism that would lead to disaster. He charged them with being selective as to what questions they were willing to examine and with stonewalling outside or contradictory evidence. Their atheistic empiricism, he declared, would cause them to become even more dogmatic than the Church they so despised and to splinter into isolated disciplines, making them suspicious of their critics and jealously protective of their particular turf. These divisions, he said, would inhibit philosophical progress and spawn a mindless fanaticism that would lead to violence. The atheists rejected Hegel and declared his attempt to discover a virtuous metaphysical explanation for man's existence to be nothing more than a veiled attempt to revive the fundamental concepts of the Christian religion. Even German disciples of the Hegelian philosophy like Marx, Engels, Bauer, Ruge, Stirner and Feuerbach were, to varying degrees, seduced by the radicals. Ironically, two of these converts from the non-violence of Hegel, Marx and Engels, later moved to Paris and collaborated in writing the *Communist Manifesto*. Loyal Hegelanists were disturbed by this fanaticism and feared that by indoctrinating uneducated workers with abstract economic and political philosophy, Marx might be creating a monster he could not control. They also feared that if rebellion brought the change from monarchical to proletarian government, unscrupulous military leaders might assume dictatorial powers. Marx conceded that in the initial stages of a revolution it might be necessary for rebel leaders to use military force to maintain order, but only until a democratically elected government could be established. Undaunted by the Hegelian criticism of his revolutionary fervor, Marx stayed the course, and his writings became the textbook for the violent atheism of the communist revolutions.

The anti-historicism that permeated academia after the rejection of Hegel kept the Empiricists from understanding the cause of the dark ages that they so longed to escape. That ignorance gave them a distorted view of the struggle between the masses and the decaying, Church-backed monarchies. Their failure to comprehend the great loss of power and knowledge that resulted from the collapse of the ancient world led them to conclude that the deep ignorance and deplorable living conditions in 17th-century Europe were the result of an evolutionary class struggle which was caused by the Church's intellectual repression of academia and the greedy policies of the capitalist monarchies. They further determined that, in the interest of human progress, this evolutionary process (later called social Darwinism) should be accelerated through revolution. Marx and Engels provided the clarion call for this workers' rebellion; their writings glorified the purity of work and heralded a day when the peasants would overthrow the monarchies, divide their wealth and justly rule the modern world. Unfortunately for the countries that adopted this ideology, the "more noble" peasant rulers of the communist revolutions have proven to be far more greedy and tyrannical than any monarchy and more ideologically repressive than the Church.

Had the history-based Hegelian philosophy prevailed, the true meaning of the collapse of the ancient world would soon have been discovered, giving philosophers a more tolerant view of the monarchs and clergy and thereby preventing much of the ideological madness that has plagued modern man. Instead, the Empiricists declared the historical accounts of interactions between the gods and man in the ancient world to be myth and, in so doing, became a cult of atheistic believers as Hegel had predicted. That catastrophic error has perverted the scientific approach to philosophy all through modern history, forcing academia to seek a new explanation for man's existence based on Natural Law. This new concept of man would be the result of a fully scientific approach, totally detached from the "primitive" paradigms of ancient history. That secular search for a scientific model of human existence had all the scholarly discipline of a California gold rush. Inspired by the freedoms unleashed in the Renaissance, atheistic scientists began throwing all kinds of speculative philosophies against the wall, hoping that something would stick. Because they had separated themselves from the historicism of Hegel, their work was largely reactionary. Humanists were desirous of discoveries that could be used to discredit the core beliefs of the Church, while scientists coveted the prestige and acclaim associated with new inventions and discoveries.

From the collapse of the ancient world to the 18th century, the single greatest force in mankind's rise to rule the world was the Church. Now the new cult of secular humanist science declared it to be an anachronism. These humanists and other intellectuals in academia hoped that religion would just fade away as new discoveries discredited its beliefs, but that was not to be. Atheistic science did hold sway in academic circles and among those who had been abused by the excesses of the Church, but the masses still bent the knee to the cults of Judaism, Christianity and Islam. As the split between the Church and secular humanist science widened, the ideological battle intensified. These philosophical machinations were strong motivating factors behind academia's rush to accept one of the most absurd hypotheses of this period, Darwin's theory of evolution.

In 1785 James Hutton published his *Theory of the Origin of the Earth*, which postulated that the earth was eons older than previously believed by Church scientists. He was also the first to enunciate the theory of Uniformitarian Principles,[20] which George Cuvier and Charles Lyell used to develop a system for dating fossils. Shortly after, in 1809, Jean Baptiste Lamarck published his *Philosophie Zoologique*. In it he introduced a classification of the species according to a "ladder of nature," which ascended from the simplest animals to the most complex. Lamarck also suggested that characteristics obtained during an organism's lifetime would be inherited by its offspring (i.e. a weight lifter's developed muscles or body scars would be passed on to his children). These highly speculative hypotheses were all components used in the work of Charles Darwin. Darwin extrapolated them out to his hypothetical conclusion that all life had evolved from a single source through the system of natural selection,[21] and he then set out to find evidence for that theory in nature. Because his hypothesis contradicted the theory of Creation, which was the very core belief of the biblical scientists, it captivated the imaginations of the atheistic secular humanists, who were only too willing to embrace a philosophy that they could use to embarrass their adversaries in the Church.

In 1859 Darwin published *The Origin of Species*, which put forth the hypothesis that man had evolved from a lower life form. The Empiricists heralded his evolutionary model of man's origins as final proof that the biblical and pagan histories were nothing more than the wild imaginings of myth-loving people in the ancient world. Standing on Darwin's unsubstantiated theory, scientists arrogantly invalidated thousands of years of recorded history in which man is universally described as a created alien who had been supernaturally placed on planet Earth by ancient gods. This, in effect, left the Empiricists without a history beyond the modern world; therefore, in order to conceal this glaring flaw in their philosophy, they claimed that it was possible to scientifically discover the origins of the planet by studying the evolutionary evidence sealed in the geology and paleontology of the earth. In an effort to squelch opposition from the Hegelian historians and the Church, they began to assert that they now had the ability to determine the true history of the planet as far back as hundreds of millions, even billions, of years and that their scientific history repudiated the eyewitness accounts of the biblical and pagan scribes. By flatly disqualifying recorded history, scientists virtually eliminated all intellectual competition. They now had a free hand to concoct a new evolutionary history because neither the Church nor the Hegelanists were capable of challenging "billions of years" of paleontological and geological evidence with recorded history that was measured in merely thousands of years. Emboldened by this intellectual dominance, the Empiricists began to fabricate a scientific history of the planet and promised to shortly produce the infamous "missing link" between man and the primates.[22] By this means they put themselves in the exulted position of the old Church hierarchy. As the early clerics had been the dictatorial arbiters of biblical history, scientists became the sole judges of evolutionary history, forcing all others to submit to their expertise. In fulfillment of Hegel's prediction, empirical science was rapidly becoming a monolithic cult that summarily dismissed all opposition to its beliefs.

With the historicism of the Hegelanists nullified, the Empiricists became more radical in their atheism, and the ideological struggle between these secular humanists and Judeo-Christianity over the origins of mankind intensified. That struggle has permeated all aspects of society and polarized our political views into intractable movements on the Left and Right. Throughout modern history, these ideologies have spawned fanatic leaders who committed unspeakable atrocities such as religion's inquisitions, pogroms, jihads and crusades, as well as the Darwinian Holocaust of Hitler's fascists and the atheistic purges of communism's Stalin and Mao Tse-Tung. This ideological mayhem has left the leaders in both science and religion afraid to openly challenge the obvious errors of the opposition, lest their own murderous history be exposed. Consequently, the argument over our origins has never been settled, leaving us confused and divided. I lay the major portion of the blame for our current malaise at the door of modern science. Unlike the theologians, who have somewhat admitted their confusion, scientists steadfastly continue to pretend to be objective while ignoring thousands of years of written history and blindly defending Darwin in the face of definitive contradictory evidence. This has locked them into a narrow evolutionary view of history, and though researchers such as Erich von Daniken and Peter Tompkins have produced volumes of evidence affirming the role of supernatural beings in the ancient world, scientists have been incapable of retracting their claim that the gods of that world are a myth. This unquestioning devotion to evolution has throttled scholarly challenges to orthodoxy within the cult of academia and forced scientists to reject all outside efforts to solve these great mysteries. As Dr. Denton revealed in his chapter "The Priority Of The Paradigm," modern scientists will cling to an erroneous position until an undeniable new reality forces them to change.

POLITICAL UPHEAVAL

As we have seen, in the early modern world the aristocracies continued to elevate the firstborn son of the royal family to the throne of power. At first, things went along as usual, but without the ability to access the power and knowledge of the ancients, these empires began to deteriorate. Economic and political power was concentrated in the hands of the aristocrats, and they used it to exploit their fellow man. This class system caused a great gap between rich and poor, and extreme poverty became commonplace. The human leaders of these monarchies knew very little about economics and political governance, so they simply taxed the poor to maintain their lavish lifestyles. Eventually, this tax burden became so onerous that it bankrupted these empires, adding starvation and disease to the troubles of the people. Isolated from this misery by position, the royal families in countries like England, France and Russia turned to the only economic policy they knew for solving such problems and imposed still heavier taxes. Many people languished in debtors' prisons, unable to pay their bills, while their families died of starvation. The weight of this oppression spawned civil unrest and, in time, led to violent revolution.

The first attempts to redress this inequity came in England in the form of the Magna Carta in 1215 and the rise of Parliamentary government later in the same century. These two events were the first successful attempts to limit the sovereignty of the nobility and give the people some influence over their governmental leaders. This process took great steps forward in the English civil war (1642–49) and the Glorious Revolution of 1688, which resulted in the supremacy of the English Parliament over the royal family. In 1776 the American Colonies broke with the Crown and founded a democratic republic. Soon after, in 1789, the French toppled their king, abolished the monarchy and established a government controlled by the business classes.

Alarmed by the early inefficiency, corruption and instability of their new republican government, landowners in France were soon longing for a return to the monarchy. In 1799 these royalists, with the support of the neighboring monarchies, set up a dictatorship with Napoleon Bonaparte as First Consul, and in 1804 they expanded his power, declaring him emperor. Napoleon wanted to create a government styled after the glorious empires of the ancient world. He sought to replace the republic with a government in which men, though denied political liberty, would have legal equality, allowing careers to be determined by talent, rather than birth. In his attempt to bring glory and prosperity to France by creating a world empire and dominating the continent, he aroused the opposition of the rest of Europe. His disastrous invasion of Russia in 1812 ended in defeat when the Austrians, Prussians, Russians and Swedes pushed him back into France. In 1814 France itself was invaded, and the emperor was forced to abdicate.

After the war, representatives of the monarchies in England, Russia, Prussia and Austria met at the Congress of Vienna (1814–15) to establish a balance of power whereby no single nation could dominate Europe. These monarchies were deathly afraid that the democratic revolutions of America and France would spread and threaten their rule. Therefore, they reinstituted the monarchy in France with Louis XVIII taking the throne, and they agreed to check the spread of such liberal ideas as nationalism, constitutional government and individual liberty. Napoleon was allowed to retain his title, but he was exiled to the island of Elba. When disputes arose among the allies at the Congress of Vienna, he escaped and made a triumphant return to Paris, where he ruled as emperor for one hundred days. After his defeat at Waterloo, he was forced to abdicate a second time and was sent as a prisoner of war to the island of St. Helena in the south Atlantic. The monarchy was once again restored.

France continued to vacillate back and forth between monarchies, dictatorships and republics for another fifty years until the third republic was formed in 1871. This political turmoil was a direct result of modern man's inability to fully sever his ties to the familiar governmental systems of the ancient world and boldly embrace modern concepts. Such indecisiveness was not peculiar to France; it was also causing problems in Germany, Italy and elsewhere. However, France is a classic example of how our fears only prolonged the agony, leading to much bloodshed. Over the years, Napoleon became a cult figure. His vision of a European empire modeled on the type of empires that ruled the continent in the ancient world would be imitated by dictators like Hitler and Mussolini well into the 20th century.

In contrast, the American experiment in the former British

Colonies was a fully democratic effort. Unlike the countries of Europe, America had no direct connection to the ancient monarchies and theocracies. As a result, its revolution was neither dictatorial, atheistic nor socialist. It did not need the blessing of the pope, nor was it the result of some frustrated leader's desire to return to the glory of a defunct ancient civilization. This was the first fully human government. Based on democracy, capitalism, individual liberty, equality and freedom from religion, this fledgling political experiment was destined to eventually lead the world. Notwithstanding, even in America, religious leaders and monarchists were continually trying to force mankind back into the physical and psychological slavery of the ancient monarchies and theocracies.

The American experiment is a shining example of the change from government by the dictatorial descendants of the ancient aristocracies to the more humane concepts of modern government. No longer would humans be subjected to the dictates of a monarch, burned at the stake by fanatics, compelled to partake in empty religious rituals or deprived of liberty and property without due process. Thanks to the innate goodness and genius of its human founders, this government would be dedicated to defending the dignity of all men. Though the fledgling America was still struggling to free itself from some of the other despicable institutions carried over from the ancient world (slavery, sexism etc.), by and large its transition to the government of man was a wonderful testimony to the human spirit.

EUROPE: AN IDEOLOGICAL CAULDRON

These early revolutions were the true beginning of our struggle to cut the psychological umbilical cord that kept us tethered to the traditions of the ancient world. As the "divinely ordained" monarchies began to fall without retribution from the gods, people realized that their leaders were merely humans with all the frailties of the masses.[23] That discovery encouraged more revolutions, and a wide variety of political ideologies began to compete for the right to run the newly developing human systems of government. As these earthly governments came to power, the ideological conflict between secular humanists and the clergy intensified. Humanists provided much of the intellectual stimulus that facilitated the development of the concepts behind these modern governments, while the Church, being deeply rooted in the ancient theocracies, supported the monarchies. The Church hierarchy believed that their adversaries were being manipulated by satanic forces and that by toppling governments blessed by the pope, they would bring down the wrath of God and possibly end the world. This skewed their perception and caused them to mistrust the motives of their ideological opponents, so they resisted these new ideas with the fervor of a Godly people warding off Satan himself. Indeed, some of the early attempts to form human governments did lead to ideological concepts that were misguided and tyrannical, but they were not satanic. We will see that though ideologies like Nazism, communism and fascism were very destructive, their inhumanity and political machinations were the result of paranoid ignorance, not demonic manipulation. Here again the failure of scientists and clergymen to recognize the implications of the collapse of the ancient world made it impossible for them to share this historic change.

With the Church and the monarchies on the defensive, secular thinkers began contending with each other, trying to discover better ways to govern. Many Christian sects had formed communities where all wealth and resources were shared according to biblical principles. These settlements, called communes, had been very successful, and their success encouraged political revolutionaries to conclude that they could use an expanded form of these closed economies to replace the economic caste system of the monarchies. In 1847 economist Karl Marx and socialist Friedrich Engels issued a manifesto stating the aims of this new secular communism. Marx also published *Das Kapital*, stating that the wealth produced by workers should go to them and that because the capitalists and monarchists would resist this change, workers must prepare for a revolution in which the capitalist monarchies would be destroyed.

Secular humanist communism was the brainchild of idealists who envisioned a Robin Hood-style government that would take from the rich and give to the poor. It was a noble but misguided attempt to right the economic and social injustices perpetrated by the monarchies against the peasants. In addition, because the Church was closely aligned with these power structures, the secular communists determined that they would have to crush Christianity and adopt the cult of Darwinian science as a state religion. In turn, the Marxist economic theories were embraced by many of the humanists as a means of further neutralizing the Church. It seems ironic, but this early Christian system of communal living would now become the economic basis for the largest atheistic and secular humanist movement in history. Nevertheless, it was doomed from the start. The humanists had mistakenly assumed that the pooling of assets had produced the economic success of the Christian communes, leading them to conclude that it could easily be duplicated in a secular environment. They failed to see that it was a fervent belief in God and the moral constraints of biblical precepts that had fostered their success. Fervent believers in any religion are very moral people among their own kind, and moral citizens will make even a flawed economic system look good. On the other hand, capitalists are beginning to realize that immorality will corrupt even the best economic system and that an honorable citizenry is an indispensable component of a sound economy. At any rate, thus began the disastrous economic jihad between the Christian monarchies and atheistic communism. That ideological conflict would keep the world in turmoil for decades to come.

WORLD WAR I

Napoleonic nationalism and the French and American revolutions had spread the idea of political democracy throughout Europe and stirred the nationalistic spirit that the ruling monarchies had tried to squelch at the Congress of Vienna in 1815. That meeting had left the German people divided into duchies, principalities and small kingdoms. Italy was similarly divided, with some parts under foreign rule, and the French and Flemish Belgians were placed under the Dutch monarchy. However, by the mid-19th century, revolutions and nationalistic fervor had succeeded in reversing much of the anti-nationalistic efforts of the congress. Belgium won its independence from the Netherlands in 1830. Germany (directed by Otto von Bismarck) became an autocratic empire in 1871. In that same year, the emperor of France was toppled, and, for the third time, a French republic was set up. All around the world, people in countries that had long been dominated by the large monarchies were seeking independence. These nationalists were not necessarily seeking a new form of government, but rather to break free of the larger monarchies and set up individual kingdoms which would independently rule their own people.

During this period, a similar upheaval was underway in Italy. Over the centuries, control of the Italian peninsula had changed hands many times as the monarchs of Austria, France, the Vatican and Spain all competed for this territory which was once the seat of the ancient Roman Empire. In the late 18th and early 19th centuries, many rulers were imposed on different areas of the peninsula by outside forces. Napoleon invaded and was welcomed by the people, who saw him as a liberator. He took Pope Pius VI prisoner and set up several republics to run the country, but counterattacks by a coalition of troops from the Austrian, British and Russian monarchies soon brought an end to this liberality. In 1805 Pope Pius VII made a deal and named Napoleon emperor. After the fall of Napoleon, King Victor Emmanuel I was the only native monarch in Italy. The country remained a battleground for the petty territorial disputes of neighboring monarchies until the rise of the Red Shirts. In 1860 a revolution took place in which red-shirted volunteers, led by the patriot Giuseppe Garibaldi, captured Sicily and Naples from Austria, and Victor Emmanuel II assumed the throne as king of a united Italy. Garibaldi eventually captured Venice and Rome, giving the pope "extraterritoriality" for the Vatican and his residence at Castel Gandolfo. This solidified Italy as an independent monarchy, though many subversive movements continued to operate within the country.

As newly independent countries like Italy began to spread their influence, tensions began to rise, spurring a vast arms buildup. Smaller kingdoms began to align themselves with the larger countries for protection against invasion and for support in clashes with their neighbors in their colonial regions. Groups that had not yet achieved independence sought the help of liberated countries as they forged national movements. This desire for independence was very prevalent among the Slavic-speaking territories, like Serbia. On June 28, 1914, a Serbian nationalist killed the archduke of Austria, Francis Ferdinand. This precipitated four years of bloody warfare with Britain, France, Russia, Japan, Serbia and, later, the United States allied against Germany, Austria and Turkey. This war was called "the Great War" because of the number of countries involved, the length of the conflict and the enormous number of casualties. The modern world had never known such cruelty. Biological and chemical warfare had been unleashed, leaving tens of thousands of soldiers and civilians maimed and diseased. In the midst of this war, Russia experienced a democratic revolution against the Czar, forcing the Russian troops to return home to defend the monarchy.

On June 28, 1919, exactly five years after the assassination, the victorious allies met at Versailles and imposed stringent war reparations on the monarchies that had lost the war. Humiliated, Germany agreed to the draconian terms and sank into economic depression. The German King and Emperor William II, who was referred to as "The Anointed Elector of the Divine Will," was the last of a long line of royal, "divinely anointed" monarchs to rule the area we call modern Germany. After World War I, he abdicated the throne, and a republic was imposed on the people by the countries that had won the war. A national constitution was adopted, and, in August 1919, Friedrich Ebert was chosen as the first president. The new Germany, torn from its ancient monarchical roots, had insurmountable problems. Many factions were vying for control of the government, and several communist uprisings had to be suppressed. The problems of changing over from the autocratic monarchies of the ancient world to the human constitutional governments of the modern era were enormous. In addition, the treaty of Versailles had imposed impossible demands on the Germans, leaving them with massive debts.

Revolution and the Rise of the "Isms"

In 1917 the Russian people overthrew Czar Nicholas II and tried to establish a democratic form of government. During this period of instability, the Bolsheviks (communists) toppled the fledgling democracy, and the first fully communist government began to rule. The leaders of the revolution took control of the country, and, as was feared by the Hegelanists, they used military force to implement their Marxist economic and social reforms. They moved quickly to make draconian changes in society, and immediately Christians, capitalists, democrats, landowners and members of the aristocracy became the targets of persecution. The Jews in Russia, who had been brutally persecuted in Christian pogroms, viewed the end of Christianity as the next best thing to the coming of the Messiah. They supported the communist takeover and, as a means of counteracting Church violence elsewhere, promoted the teachings of Marx with their brethren throughout the world. Russian Jews had been so brutalized by the Christians that they found forced atheism more palatable than forced baptism. The fact that Marx was a fellow Jew also helped ease their otherwise illogical infatuation with atheistic science.

When Joseph Stalin succeeded the founders of the revolution in 1924, the full impact of Marx's dream of peasant rule was realized. Stalin's real name was Iosif Vissarionovich Dzhugashvili; he was the son of a factory worker who had died of wounds received in a drunken brawl. His mother, Yekaterina, did housework for local people to keep the family together. A deeply religious woman, she worked to get young Iosif into the priesthood as a means of advancement. She managed to secure a place for him in the local Russian Orthodox religious school, and his grades won him a scholarship to the theological seminary. It was there that he became a member of the secret society that formed the genesis of the Bolshevik revolution. That association caused him to be expelled from the seminary shortly before graduation, and he then became a journalist for the radical underground, using the pseudonym Stalin, or man of steel, as his byline. He was arrested and exiled to Siberia several times, but always managed to escape. Stalin worked his way up through the ranks of the Bolsheviks, eventually becoming a protégé of Lenin. In 1919 he was made head of the Workers' and Peasants' Inspectorate. This agency had the power to investigate any official in the country. It was from this post that he began to eliminate his rivals and consolidate a political power base. In 1921 he helped Lenin crush an open revolt among the peasants, workers and sailors, and used ruthless military force to recapture the independent state of Georgia. In 1922 he was appointed to the new post of Secretary General, making him head of the entire party apparatus. This placed him in a strategic position to win the struggle for leadership after the death of Lenin. In 1924 Stalin craftily used his allies in the party, Kamenev and Zinoviev, to oust Lenin's chosen successor, Leon Trotsky, and later, Kamenev and Zinoviev themselves were dispatched. Stalin quickly tightened his grip on power, banning criticism within the party, abolishing freedom of expression and association, and centralizing the bureaucracy. Each of these moves was in direct contradiction to the original aims of the founders of the revolution. In the early 1930s, Stalin's repressive measures led to a secret plot to replace him with Sergei Kirov, a party leader in Leningrad. Stalin learned of the plot and had him murdered. He then began a purge of terror that lasted more than five years, in which he executed virtually all of the party and military leaders, and incarcerated, murdered or exiled multiplied millions of Soviet citizens. With the help of the secret police, Stalin became the sole dictator over both the party and the country. In a way that he himself could never have foreseen, Marx's dream of peasant rule had now become a reality. Unfortunately, the peasant dictator, Josef Stalin, would prove to be more destructive and murderous than any pope, king, emperor or czar who had ruled in the modern world. As feared by the Hegelanists, Marxist revolutions were always followed by military dictatorships that were ruled by paranoid peasants. No exact figure has been established for the number of peopled killed by Stalin, but estimates run as high as fifty million.

Under Stalin's iron rule, communism's survival seemed assured. In fact, across Western Europe, socialism, a less radical form of communism, appeared to be the wave of the future. Inside Russia, Stalin brutally crushed all opposition. He closed the churches, confiscated their wealth and shipped the priests to Siberia for "retraining." Declaring support for a worldwide workers' revolution, Stalin encouraged communist rebellions in many countries, supplying local insurgents with money, arms and training in guerrilla warfare. In addition, subversives were trained to infiltrate major institutions such as labor unions, universities, churches, the press and political parties in order to destabilize countries from within. Christianity saw this atheistic revolution as the greatest threat to its power in the history of the Church. As a result, the Vatican looked to align itself with any government that would help destroy this "evil" system before it could spread throughout the world.

ITALY

In Europe after WWI, nationalistic fervor once again stirred a desire to throw off the remaining monarchies. The worldwide communist movement was very active in Italy, and a conservative nationalist journalist by the name of Benito Mussolini started a movement to resist their efforts and strengthen the national government. In 1919 Mussolini organized the *Fascismo*. The term *fascist* was taken from the name of the emblem carried by ancient Roman magistrates as a symbol of their authority.[24] Mussolini, like Napoleon, dreamed of restoring the power and majesty of the ancient Roman Empire. In 1922, fearing that the communists would topple the weak Italian monarchy, he sent a column of his followers to march on Rome. King Victor Emmanuel III, intimidated by his show of strength, named him premier, and Mussolini immediately embarked on a campaign of empire-building. Thus was born the notorious Fascist movement.

By the 1930s, Europe was a breeding ground for all sorts of ideological conflicts. Atheistic communism, Christianity, Islam, Judaism, fascism, democracy, nationalism, capitalism and secular humanist science were all vying for power. Christianity and the monarchies were trying to maintain the status quo, while the emerging ideological forces were flailing about in the darkness, hoping to develop a new system of government with which to topple the remaining monarchies and theocracies. This intense ideological competition caused the leaders of these movements to mistrust the motives of their rivals; consequently, all were secretly plotting against each other. In the midst of this cacophony of ideas, capitalism suffered the Great Depression, the Church continued to splinter, communism was pushing a worldwide revolution and scientists were trying to convince man that he was an ape.

GERMANY

As part of this growing ideological storm, a charismatic ruler emerged in Germany. Adolph Hitler rose to power at a time when numerous forces within the country were demanding change. Hitler used this confusion to his benefit, and under the banner of the National Socialist Party, he promised to restore order and return the country to the magnificence the German people had manifested in the glorious civilization of their ancestors, the ancient Goths. With adroit political moves, his party won a majority of the seats in the Reichstag, forcing President Marshal von Hindenburg to make him Chancellor. Hitler was well aware of how the Bolsheviks had taken control of the revolution in Russia; therefore, he quickly moved to eliminate all political opposition and, in 1938, named himself dictator for life. Like Napoleon in France and Mussolini in Italy, Hitler excused this tyrannical use of power as a necessary evil. He claimed it would enable him to resist a takeover by the international communist movement and guarantee the independence of the German people. The economy had been so devastated by World War I and the Great Depression that people were starving in the streets. Appealing to their great heritage, Hitler reminded his countrymen of how their ancestors had withstood the assaults of the mighty Roman Empire, eventually bringing it to its knees. The message was irresistible, and his call to arms rekindled a sense of pride in the broken hearts of the humiliated, defeated and financially destitute German people. While Russia's atheistic communism and America's democratic republic were attempts to sever ties to the ancient world, fascism and nazism, like the Napoleonic empire, were attempts to return to its power, glory and leader worship. As a result, the validity of these fascist-style governments was measured by the number of people who could be duped into following a charismatic leader, rather than the voluntary intellectual assent of a people embracing a benevolent new ideology.

In 1934 Hitler began to rearm Germany and, in a quick move, retook possession of the Rhineland. These actions were direct violations of the Treaty of Versailles. Two years later, he refused to pay any further reparations for World War I and began to rattle his saber at the rest of Europe. When Germany strengthened its alliance with Italy and Japan, things started to look ominous. Even countries as remote from the growing crisis as America were becoming nervous at the prospect of a worldwide conflict. As the war clouds gathered, advocates of mainline religious ideologies were forced to find military allies. Some of these alliances can only be described as bizarre. The Jews stood with the atheistic communists; the Vatican, with the quasi-humanist and pagan fascists, while science was happy designing the weapons of war for both sides.

To avoid having to fight a war on two fronts, which had caused Germany's defeat in WWI, Hitler signed a peace treaty with Russia. This left the traditional allies divided, and the Axis forces began to invade the weaker countries in Europe and Africa. France and Britain had an alliance, but internal unrest in France and a weak government in Britain delayed their reaction, allowing the Axis to gain a quick advantage. They rapidly spread across Europe and down into North Africa. When the western front had been pushed to the English Channel, Hitler scrapped his ten-year non-aggression treaty with Russia and attacked along the eastern front. Stalin, fearing that Russian Christians wouldn't fight to defend atheistic communism against an enemy supported by the Vatican, offered a desperate deal to the former leaders of the Russian Orthodox Church: support him in his battle against the Nazis, and he would restore their wealth and power. The Russian priests made this "deal with the devil" and, from reopened churches, began to encourage the Russian people to defend Mother Russia and save Stalin. We now had the ridiculous spectacle of Russian Christians fighting to save atheistic communism from the Nazis and their allies in Italy and the Vatican.

JAPAN

At a time when the Western monarchies were dividing up much of Asia, Japan managed to remain independent and became a world power by modernization. Japanese leaders took many of the revolutionary Western ideas and molded them to their particular situation. They studied the European and American constitutions and, in 1889, modeled one of their own, with a parliament called the *Diet*. This liberalization did not effect the emperor. He continued to be worshiped as a god and had the power to issue laws through personal edicts and to adjourn parliament by decree.

The Japanese saw the Russian forces in China as a threat to their dominance of Asia. After negotiations failed to get them to withdraw, in 1904 they launched a surprise attack on the Russian Naval vessels berthed at Port Arthur in China. The Japanese won the war, gaining great prestige as a power to be respected. In 1910 they annexed Korea, and they set up a puppet government in Manchuria in 1932. Though Japan had supported the allies against Germany during World War I, politics in Asia had changed dramatically by the 1930s. When the Japanese army invaded northern China in 1937, they received the support of Germany and Italy. Feeling threatened by the subversive nature of the revolution in Russia, Japan welcomed this offer and aligned itself with the anti-communist stance of the European fascists. This was a natural alliance, as the ancient-world imperialism of Japan harmonized with the style and aspirations of the fascist dictatorships. The worship of the Japanese emperor as a god was also in tune with the adulation afforded Hitler and Mussolini as they tried to affect the demigod image of the ancient rulers. In turn, the Japanese leaders felt comfortable with the dictatorial nature of the Italian and German regimes. Their alliance, called the Axis, was formidable, and by the late 1930s, what was rapidly becoming a worldwide ideological cauldron was ready to erupt into the ugliest conflict in modern history. When the Americans opposed Japan's aggressive posture in the Pacific and used an oil embargo to show its displeasure, the Japanese tried to repeat the strategy that had been so successful against the Russians in China and began to plan their attack on the U.S. Naval fleet at Peal Harbor in Hawaii.

CHINA

In China, inspired by the innovations of Europe, a similar nationalistic spirit was taking hold. In 1911 a revolution headed by Sun Yat-Sen toppled the Manchu dynasty and set up a republic. Unfortunately, any hope of real change was dashed when the first president, Yuan Shi-Kai, tried to reinstitute a dynasty. His efforts failed, and the country fell into the hands of the warlords, who took control of their individual provinces, leaving China with no central government. The warlords supported the allies in WWI, and at the peace conference at Versailles, they asked to have all foreign powers removed from their territories. The request was denied. In 1926 nationalist leader Chiang Kai-Shek rallied a coalition of forces (including the communists) and took control of the country. He drove all foreign forces from China and, fearing a counter-revolution, turned against his communist allies. Though Chiang's government was nationalistic, it was also autocratic; therefore, the communists were very successful in convincing the people that they were the only movement advocating a complete break with the ancient dynastic system and the creation of a truly modern government. Years later, as had happened in Russia, the communists, led by Mao Tse-Tung, did topple the nationalists, and the leaders of the original revolution were forced to flee. In 1948 Chiang escaped to Taiwan with his forces, and thus began the two Chinas.

World War II and Hitler's Darwinian Holocaust

With the Axis forces on the march in Europe, the United States resupplied Russia and Britain, and built up its forces in the Pacific as a counterbalance to Japan. The Japanese saw this as a threat and finalized their preparations for a surprise attack on the U.S. Naval base at Pearl Harbor. With the United States' entry into the war, mankind's fears had finally brought him to a truly global conflict. There would be no winners in this debacle.

Due to Christianity's history of persecuting the Jews, the terrible tragedy that befell them during WWII has been mischaracterized as something that resulted from Christianized Europe's rampant anti-Semitism. In fact, though six million Jews were killed in this disaster, at least eight million non-Jews were similarly eliminated. In order to avoid the confusion that results when it is referred to as the Jewish Holocaust, I have deliberately renamed this tragedy. Separating the Jews from the rest of the victims of this Darwinian debacle allows us to blame elusive culprits like hate, thereby obscuring the terrible implications for science. Though anti-Semitism was indeed a factor in the Final Solution, it was not the cause of the Holocaust. A few hate-filled fanatics could never have implemented such devastation. The decision to systematically eliminate fourteen million people required not only political will, but the kind of legitimacy that can only be afforded by an ethical or scientific authority. The Holocaust resulted from a governmentally authorized attempt to use Darwinian philosophy to manage a futuristic civilization in the Third Reich. It was a scientifically planned effort to create a master race of pure-blooded Aryans through the implementation of Darwin's theories on artificial, or methodical, selection. The fact that European Jews, as a group, took the largest number of casualties has muddied the waters, allowing atheistic science to escape blame for this genocidal experiment. It is time to set the record straight.

Most people can readily see the connection between the perverted doctrines of early Christianity and Europe's rampant anti-Semitism. For centuries, in purges, pogroms and witch hunts, these mistaken beliefs caused hundreds of thousands of innocent people to be killed. Yet even under the Christian Inquisition, each heretic, witch or Jew had to be given a trial and executed on an individual basis. If a Jew recanted and submitted to baptism, his life would be spared. When these mechanisms failed to solve the "Jewish Problems" of England and Spain, they simply expelled the entire population from the country. These were terrible tragedies, but they pale when compared to the genocide caused by the doctrines of amoral science. Unlike the Inquisition, the Darwinian Holocaust was not about religious beliefs; consequently, there was no chance to recant nor any baptism to cleanse one of the imagined genetic defects that scientists considered offensive. When it was over, the appalling death statistics from the Inquisition would seem paltry next to the multiplied millions slaughtered by the scientific doctrines of social Darwinism.

During the years when Hitler was forming his political ideology, Darwin was the darling of modern science. Scholars, revolutionaries, scientists and philosophers were falling over themselves trying to prove their intellectual prowess by expounding on the virtues of his theory of evolution. Not surprisingly, a persistent theme throughout Hitler's book, *Mein Kampf*, is social Darwinism—the concept that human civilizations are subject to the evolutionary law of the survival of the fittest. Darwin's theory led Hitler to believe that the very existence of the German people was being genetically threatened from within by intermarriage with foreigners and politically threatened from without by Marxism, which, as he saw it, was secretly being funded and manipulated by the Jews.[25]

Darwin's theory that science could improve the human race through artificial selection[26] gave rise to the Nazi concept of creating a master race. Taking this theory to its logical conclusion, Hitler's scientists planned to systematically nurture the Aryan race. First, they would protect it from further defilement through intermarriage with what Darwin had called the savage races.

Second, they would eliminate the defects that jeopardized the stability of the Aryan gene pool, such as mental illness, physical weakness, homosexuality and criminality. Third, they planned to wipe out the threat posed by rival races within the Fatherland. Darwin's doctrine mitigated the need for scientists to treat people as individual human beings, classifying them instead as groups along an evolutionary chain. It allowed Nazi scientists to declare large numbers of people to be either members of a rival race or individual mutations that threatened the development of a healthy Aryan society in Germany. Now any individuals or groups deemed inferior by Darwinian scientists in the Third Reich could simply be eliminated en masse through clinical-sounding "artificial selection experiments" aimed at perfecting the quality of the master race.

In the following quote from *Descent Of Man*, Darwin reveals how his ignorance of the true cause of the collapse of the ancient world contributed to his mistaken theories.

> All that we know about savages, or may infer from
> traditions and from old monuments, the history of
> which is quite forgotten by present inhabitants, shew
> that from the remotest times successful tribes have
> supplanted other tribes. Relics of extinct or forgotten
> tribes have been discovered throughout the civilized
> regions of the earth, on the wild plains of America,
> and on the isolated islands in the Pacific Ocean. At
> present day civilized nations are everywhere
> supplanting barbarous nations, excepting where the
> climate opposes a deadly barrier; and they succeed
> mainly, though not exclusively, through their arts,
> which are the products of the intellect. It is, therefore,
> highly probable that with mankind the intellectual
> faculties have been mainly and gradually perfected
> through natural selection; and this conclusion is
> sufficient for our purpose. . . .
> (*Descent of Man*, Ch. 5 p. 160)

This is a perfect example of how anti-historicism and the

resultant ignorance of the true cause of the collapse of the ancient world has led atheistic scientists to reach one erroneous conclusion after another. In this case it causes Darwin to assume that the transition from ancient to modern civilization was the result of evolution and natural selection. This is a huge mistake that perverts his entire outlook. He wrongly concludes that the ancient civilizations were gradually supplanted by races with superior morals and intellect. As a result, he feels comfortable with the idea that the advanced civilizations of Europe should conquer the "barbarous nations" of the world. He not only sees this as inevitable, but as absolutely essential to the survival of the species, since, in his view, human racial groups are bound by the law of the of the survival of the fittest. That tragic mistake allowed Nazi scientists to be clinically detached while using mass sterilization and extermination as instruments of methodical selection in the Holocaust.

Another quote from his writings provides some insight into the racism of Darwin. Here he refers to his less affluent fellow humans as "savages" and suggests that we sterilize the weaker members of civilized society.

> With savages, the weak in body or mind are soon eliminated; and those that survive commonly exhibit a vigorous state of health. We civilized men, on the other hand, do our utmost to check the process of elimination; we build asylums for the imbecile, the maimed, and the sick; we institute poor-laws; and our medical men exert their utmost skill to save the life of every one to the last moment. There is reason to believe that vaccination has preserved thousands, who from a weak constitution would formerly have succumbed to small-pox. Thus the weak members of civilised societies propagate their kind. No one who has attended to the breeding of domestic [farm] animals will doubt that this must be highly injurious to the race of man. It is surprising how soon a want of care, or care wrongly directed, leads to the degeneration of a domestic race; but excepting in the

case of man himself, hardly any one is so ignorant as to
allow his worst animals to breed.
(*Descent of Man*, Ch. 5 p. 168)

This is the classic Darwinian approach to his fellow man. He
sees himself in the role of a scientific cattle rancher, attempting to
purge objectionable defects from the human herd. His theory that
man evolved from the primates allows him to view the human race
as merely another breed of animal that desperately needs to be
thinned in order to ensure its survival. Darwin lampoons all of
mankind's humane efforts to sustain his fellow man as ill-advised
methodical selection that is thwarting the benefits which would
flow from a system of natural selection, as it exists in the "savage
races." He warns that if we continue to use methodical selection in
this sympathetic manner, the bloodlines of the civilized white races
will be weakened from within and the physically superior savage
races will take over the earth. Darwin goes on to suggest that we
should instead use methodical selection to prevent the less desirable
members of the civilized herds from procreating, thereby ensuring
civilization's survival. His paranoid prediction was not lost on Nazi
scientists. It allowed them to assuage their natural sympathies while
slaughtering fellow humans they considered defective for society's
greater good. These racist philosophies are at the very heart of the
Darwinian Holocaust.

His next statement is even more disturbing.

> The aid which we feel impelled to give to the helpless
> is mainly an incidental result of the instinct of
> sympathy, which was originally acquired as part of the
> social instincts, but subsequently rendered more tender
> and more widely diffused. Nor could we check our
> sympathy, even at the urging of hard reason, without
> deterioration in the noblest part of our nature. The
> surgeon may harden himself whilst performing an
> operation, for he knows that he is acting for the good
> of his patient; but if we were intentionally to neglect
> the weak and helpless, it could only be for a contingent

benefit, with an overwhelming present evil.
(*Descent of Man*, Ch. 5 p. 168–9)

Hitler's scientists used this concept to justify eliminating the overwhelming present evil of genetically inferior humans from the German herd for the contingent benefit of a pure and healthy mankind in a perfect Fatherland. They even went so far as to parade pure-blooded Germans through the streets as traitors for having intermarried or shared a bed with a Jew or Gypsy.

The fact that Darwin singled out Jews and Gypsies for special attention in his writings also provided a certain scientific legitimacy to the work of the Nazis.

> The uniform appearance in various parts of the world
> of gypsies and Jews, though the uniformity of the latter
> has been somewhat exaggerated, is likewise an
> argument on the same side.
> (*Descent of Man*, Ch. 7 p. 242)

In this quote, Darwin makes reference to the similar appearance of Jews and Gypsies when comparing the methodical cross-breeding of farm animals with how natural selection creates skin color in humans. The very fact that Jews and Gypsies didn't normally intermarry with the indigenous populations made them perfect guinea pigs for Darwin, and it is all too clear that his devotees in Nazi Germany found them equally compelling and saw them as the very fulfillment of his ominous predictions concerning the danger posed by the savage races.

This was the type of ideological madness that was driving the agendas of the emerging leaders of Europe as the monarchies began to fall. Hitler concluded that if he was to win the race to lead the modern world, he would first have to restore the German economy and create an orderly society. Most modern writers portray Hitler as some kind of maniacal monster who enjoyed killing people he hated. Nothing could be further from the truth. Hitler saw himself as a futuristic leader with a grand vision for a new socialist government that would restore order to a chaotic world. He envisioned a Napoleonic-style empire that would utilize the brightest minds of the modern world to bring mankind health and prosperity through a master race that would rule the earth. Social Darwinism had convinced the Nazi scientists that civilized ancient societies like Rome and Greece had been destroyed from within. They theorized that intermarriage with immigrants and rival races had weakened these empires and led to their downfall. Hitler planned to avoid that mistake by combining the glory, power and leader worship of those ancient civilizations with Darwinian science to create a modern government that would reign for a thousand years. The Nazis planned to fight a relentless struggle against their internal and external enemies and, in the process, conquer enough territory to protect the Fatherland; a race of pure-blooded Aryans in a new socialist Germany would be the crown jewel of this great empire. To implement this grandiose scheme, Nazi scientists planned to end the mixing of bloodlines that had resulted from immigration. A totally pure Aryan race would be isolated from any integration with foreigners, and even Germans who had intermarried with outsiders would have to be excluded. These disciples of Darwin saw the Nordic-Aryan race as the top of the evolutionary chain. In an effort to implement the first full-scale test of Darwin's theories on methodical selection, they set about eliminating all the "defective" and non-Aryan members of the population. They ordered the mass sterilization of over four hundred thousand mixed-blooded German citizens and identified Jews, Gypsies, Jehovah's Witnesses, criminals, communists, homosexuals, dissidents and all handicapped people for special attention. The less easily identifiable groups, like Jews and homosexuals, were forced to wear armbands[27] in order to eliminate the possibility of accidental sexual encounters with pure Aryans.

THE JEWISH PROBLEM

By the late 1930s, the xenophobia produced by social Darwinism had reached its peak. To Nazi scientists, the growing populations of "savage races" in Europe appeared to be an exact fulfillment of Darwin's ominous predictions. They began to see them as a dagger ready to pierce the heart of the more civilized white races of Europe. The Jews, in particular, were not only a growing percentage of European civilization, but they controlled a large portion of the German economy. Pure-blooded Aryans were already serving as maids, nannies and day laborers for the more wealthy members of this "savage race." Hitler's Christian upbringing, combined with the fact that the Jews had supported the rise of communism in Russia, led him to see them as a fifth column. He considered them to be a rival civilization within the borders of Germany who were responsible for everything from the death of the Christ to the economic troubles of the Great Depression. Marxism, the only viable opposition to Nazism, was very popular among the Jews; therefore, as the largest non-Aryan group in Germany, they posed a threat not just genetically, but politically as well. Their vast numbers, great wealth and Marxist ideology posed a far more ominous danger than all the other non-Aryan or defective members of German society combined. To eliminate their potentially dangerous ideological competition and avoid further contamination of Aryan bloodlines, all Jews were immediately herded into ghettos like quarantined cattle. An indication of how gravely they perceived the problem can be seen in the fact that resources vital to the war effort were instead dedicated to finding a "Final Solution" to this impending Darwinian disaster.

A SCIENTIFIC UTOPIA

In the early stages of the Holocaust, scientists invested a huge effort in an attempt to categorize all non-Aryans by physical characteristics (i.e. eye and skin color, nose and head shape) in order to determine the genetic differences between the savage races and pure Germans. During this period their goal was to try to Aryanize some of these people, in the hope of making them useful citizens. Aside from being ghastly, their early experiments reveal the Darwinian nature of their efforts. Scientists injected blue dye into the eyes of Jews and Gypsies as part of an experiment in color control, and similar attempts were made to manipulate skin and hair coloring. When their attempts to medically correct these "defects" failed, scientists began to use people like lab mice in more brutal military experiments. They dropped them into ice water to see how long a downed pilot could survive in Arctic waters and subjected others to radical changes in atmospheric pressure until their ear drums burst. When these poor people served no further purpose as guinea pigs, the scientists merely gassed them and cremated their remains. Later, as more resources had to be allocated to the external war, scientists began to realize that they were running out of time to achieve their Darwinian utopia and moved to simply eliminate these people as quickly as possible. At that point, the incredible paranoia generated by this strange mix of ideologies, known as Nazism, quickly erupted into what was eventually called the Holocaust. Hitler ordered the far-reaching hands of the Nazis and their allies to round up the rival races, along with all the other defective members of society, and ship them to concentration camps for extermination. In all, more than fourteen million human beings—men, women and children—were brutally sacrificed on the altar of experimental science, the first governmentally sanctioned attempt to implement Darwin's theory of methodical selection.

Another quote from Darwin may help us to understand the scientific concepts that motivated apparently normal and intelligent human beings to inaugurate such a perverted scientific experiment.

> . . .Man differs widely from any strictly domesticated animal; for his breeding has never long been controlled, either by methodical or unconscious

selection. No race or body of men has been so
completely subjugated by other men, as that certain
individuals should be preserved, and thus
unconsciously selected, from somehow excelling in
utility to their masters. Nor have certain male and
female individuals been intentionally picked out and
matched, except in the well-known case of the
Prussian grenadiers; and in this case man obeyed, as
might have been expected, the law of methodical
selection; for it is asserted that many tall men were
reared in the villages inhabited by the grenadiers and
their tall wives. (*Descent of Man*, Ch. 4 p .112)

Here, Darwin laments the fact that no race has ever been
sufficiently subjugated by other men to allow a true test of
methodical selection as a means of creating a superior breed of
human, and he regrets that the experiments in Prussia weren't
scientific enough to reach a conclusion. Should we be surprised
that his followers in Nazi Germany tried to fulfill his dream of a
full-scale, "truly scientific" methodical selection experiment? Nazi
scientists, steeped in the amoral concepts of atheistic science, acted
like mind-numbed robots. They felt no pang of conscience when
carrying out the methodical extermination of Jews, Gypsies and
even fellow Germans for the contingent benefit of creating a
Darwinian master race in a perfect Fatherland.

Current scientists try to salvage the loathsome doctrines of
methodical selection and evolution by claiming that Nazi scientists
perverted or misunderstood Darwin. This is the equivalent of
modern Christians claiming that the people who put the Jews on
the trains weren't "true" Christians or that the perpetrators of the
Inquisition misinterpreted Church teaching. Sooner or later, we
will all have to face the fact that these perverted doctrines of science
and religion have caused otherwise good people to do terrible
things. Nazi scientists didn't misunderstand Darwin; to the
contrary, they did their best to carry out his loathsome doctrines
with characteristic German precision. If the cult of atheistic science

continues to glorify Darwin and teach his perverted doctrines to its converts, these new adherents will eventually repeat this madness under the guise of genetic manipulation or cloning "for the good of society." If Darwin is right, then another Holocaust is not only inevitable, but scientifically justifiable. It is my position that he is dead wrong; therefore, scientists should repudiate his teachings, drive them from our elementary schools and apologize to humanity for their part in the Holocaust.

After I appeared on a radio talk show in New York City, a listener sent me a transcript of a speech by Kalev Pehme. Mr. Pehme teaches on the subject of racism, and he lays the blame for the Holocaust squarely at the door of Darwinian science. In his lecture "The Myth of Race" delivered at the Henry George School for Social Sciences in New York on February 4, 1994, Pehme stated:

> . . .Darwin is considered something of a free thinker
> and a hero who stood against the superstitious,
> particularly the fundamentalist Christians. While he
> indeed did that, the content of what he taught is
> greatly ignored and merely accepted.

I have found that science teachers will defend Darwinian philosophy to the death because they think it disqualifies the "Creation myth." Yet, when pressed, they grudgingly admit that they have only skimmed his books. Nazi scientists like Dr. Joseph Mengele are held in the utmost contempt throughout the world, yet the erroneous hypotheses of their mentor, Darwin, continue to be promulgated in every school in this country, ironically, oft times by Jewish teachers.

Pehme further stated that,

> Darwin's theory of evolution is nothing more than a
> political statement beginning with *The Origin of Species*
> and culminating in his *Descent of Man*. While Darwin
> parades as a natural scientist, he is nothing more than a
> political thinker who believes that the wealthy whites

are superior to the poor; that whites are superior to all
other races; and that it is inevitable that white men will
eventually conquer the earth and, by implication, kill
all inferiors.

I agree! Darwin's writings are decidedly racist, and they were
the scientific catalyst behind the three most destructive philosophies
of the twentieth century: communism, fascism and Nazism. All of
those ideologies have since been discredited, but Darwin still lives.
His perverted theories remain the cornerstone of modern science,
and they are required course studies for every Junior High School
student in the Western world. We wouldn't dream of allowing the
corrupt concepts of Hitler, Mussolini or Stalin to be foisted on our
schoolchildren, yet their mentor is portrayed as a great scholar, the
revered icon of academia. How can it be that we have allowed this
biased philosopher to continue unchallenged through all of these
years? The answer is that scientists have become so powerful that
no one dares to take them on.

Why do scientists and academicians feel compelled to cling to
the flawed and racist doctrines of Darwin with such tenacity? The
fanaticism and vindictiveness of their personal attacks on dissidents
like Dr. Michael Denton remind one of the Church's attacks on the
opponents of the geocentric solar system. Given a moment's
thought, the reason for their obsessive preoccupation with
evolution becomes obvious. If the Darwinian myth is exposed,
there's no place for scientists to turn but the Bible. If Darwin falls,
it's certain that Creation must be the true explanation for the
origins of man. To scientists, the fall of Darwin would be the
equivalent of what the fall of Jerusalem was to the theologians—a
truth too awful to face. So they protect him at all costs and try to
divert attention from the fact that the racist doctrines of their
atheistic messiah were the root cause of the Holocaust. Their cult-
like devotion to Darwin is preventing scientists from being candid
about the role their atheistic beliefs played in this genocidal
experiment. If we are to alleviate the paranoia that the Holocaust

has left on the human psyche, we will have to identify all of the ideological cults that caused it. Not just fascism and Nazism, but the cults of Christianity, Judaism, secular humanism and Darwinian science as well. Christianity's charge that the Jews had killed the Christ mollified the consciences of local officials in Christian countries like Poland and France, who, after confiscating their possessions, crammed the Jews into box cars for shipment to concentration camps. It may be added that Jewish ideology also played a role in this tragedy. By requiring its adherents to live a separated lifestyle and declaring them to be the chosen people of God, it was, religiously speaking, poking a finger in the eye of its Christian neighbors. We must stop using Hitler and the Nazis or "hate" as scapegoats and blame the failure of our ancestors to face the awful truth inherent in the fall of Jerusalem. That tragic denial is ultimately the root cause of not only the Darwinian Holocaust, but all the other jihads, crusades and holy wars of modern history as well.

THE COLD WAR

When World War II ended, many countries were on the verge of economic collapse. The atomic bomb had ended the war in Asia, but fear of its terrible power would plague mankind for the foreseeable future. Fascism was dead, except in the paranoid minds of the Jews who had survived the Holocaust and the Marxists who had survived the invasion of Russia. They saw it lurking behind every non-communist tree. Communism and capitalism emerged as the dominant economic forces, and the ideological battle lines were already being drawn. Non-communist countries were becoming increasingly concerned about the threat posed by this growing workers' revolution. What had started as a means of overthrowing the ancient monarchies had now degenerated into a fanatic ideology bent on taking over the world. The communists were even attempting to overthrow modern democracies and republics that had been rightfully elected by the people. Marxist insurgents were found trying to destabilize the governments of Britain, France and the United States. In response, powerful figures in America and Britain began to suggest that the allies go into Russia, topple the communists and install a democracy. This, combined with the fact that the United States had emerged from the war as a military and economic giant, engendered much paranoia in the emerging Soviet Union. Perceiving this threat, Stalin annexed the Baltic States and much of Eastern Europe as a buffer zone against invasion, isolating them and Russia proper from Western ideas with policies that came to be known as the "Iron Curtain."

With Hitler out of the picture, Stalin renewed his persecution of the Church in countries like Poland, Czechoslovakia and Hungary. Subjecting these Catholic countries to the atheistic dictates of communist regimes exacerbated the growing tensions between East and West. This crackdown moved the Vatican, whose economic teachings were more in line with socialism, to seek help from the capitalist West. The Church's support for dissidents in Eastern Europe led to the Hungarian revolution of 1956, which was brutally crushed with Russian military might. Thus the topsy turvy ideological cauldron continued to boil as though the second World War had never occurred. The bordering countries in Western Europe perceived the new Soviet Union as a power to be respected and, fearing a communist takeover, tried to accommodate this powerful neighbor by being amenable to socialism. The Warsaw Pact and NATO military alliances drew the dividing lines, and the world settled into a new ideological conflict, called the Cold War. The Russians, who had been our allies in World War II, quickly became our worst enemies solely because of these confused and unsubstantiated ideological beliefs.

THE UNITED NATIONS AND THE STATE OF ISRAEL

With the end of the war, it was time for the world community to deal with the plight of the Jews. Jewish refugees who had survived the camps had lost all their possessions, and their pre-war land rights had been given to local Christians as rewards for their capture, so they had little hope of returning to their homes. This was a hot potato. Everyone acknowledged that a terrible crime had been perpetrated against the Jews, but no one knew what to do about it. Here was a group of people who, though they had no country of their own, thought of themselves as a separate nation. They had been brought up to believe that they were still a chosen people who would be returned to Israel when the Messiah came. After hundreds of years of persecution, pogroms and now the Holocaust, the Zionists were no longer willing to wait for the Messiah. They decided to turn to the United Nations as a means of returning to Israel, whether God was ready or not. Many orthodox rabbis stood in opposition to these Zionist goals, claiming that the Scriptures supported their position that only God could restore the Jews to the promised land.

Unwilling to address the obvious, yet all-too-volatile, theological aspects of the problem, world leaders made a mistake not unlike that of the Nazis. As Nazi scientists had deceived themselves by pretending that the religious dilemma caused by a civilization of Jews living a separated existence in a Christian country would lend itself to a scientific solution, world leaders now decided to pretend that it would lend itself to a political solution. To avoid the obvious political and theological implications, Hitler's Jewish Problem was euphemistically referred to as the "plight of the Jews." However, it was quickly discovered that anti-Semitism was more than just a Nazi quirk. In fact, most Christian and Islamic countries were unwilling to have anything to do with these refugees. Judaism's claim that Jesus was not the Messiah was perceived as an insult by both Christians, who believed he was the son of God, and Muslims, who believed he was a great prophet.

This made it extremely difficult to find countries that would be willing to accept the Jews. Had the United Nations candidly addressed that problem, much could have been achieved. Instead, they tried to craft a political band-aid that would be capable of covering this open and festering religious wound. Understandably, sympathy for the plight of the Jews made an objective assessment of the situation impossible. Anyone who suggested a dispassionate approach to the problem was branded an anti-Semite, while Muslims and Christians, even those who had no part in the Holocaust, were made to feel guilty.

Theologically unqualified and politically unwilling to address the contentious religious beliefs that are the root cause of this philosophical problem, the U.N. tried to find a remedy for the symptoms. World leaders moved to solve the fact that the Jews were hated by their fellow citizens in most Christian and Muslim countries by creating a separate homeland for them in Palestine. The Zionists had been lobbying for a Jewish state in the Middle East for decades. At the time, Palestine was a British territory, and, as early as 1917, A. J. Balfour had declared Britain's support for the proposal. Disregarding the fact that a previous influx of Jews fleeing the anti-Semitism of World War I had caused great rancor among the Muslims there, the United Nations adopted a plan to create the Jewish state. The same allies who had recently called Mussolini and Hitler madmen for attempting to recreate the ancient empires of the Romans and Goths had now committed the Jews to the equally impossible task of recreating ancient Israel. Tragically, like the Christian Crusaders of a thousand years earlier, the Jews would soon discover that wresting control of that ancient land from its modern Muslim inhabitants is a deadly pursuit.

In 1948 the Zionist dream of a homeland for the Jews became a reality. However, by failing to address the obvious religious problems first, the United Nations subjected not only the already beleaguered Jews, but the Muslims as well to decades of holy wars, terrorism and bloodshed. Before the ink could dry on the agreement, the first of many wars erupted. Political leaders and the media ignored the obvious and euphemistically referred to these religious battles as Arab-Israeli conflicts, blaming "hate" for causing them. In reality, they were yet another replay of the endless crusades, jihads and holy wars to rescue Jerusalem from the "infidels." We will continue to relive these ancient horrors until we muster the courage to candidly address the reality inherent in the fall of Jerusalem in A.D. 70.

Now the Jews had their homeland, but many didn't want to go to a hostile land and try to resurrect an ancient civilization in the modern world. The more idealistic Jews fleeing Europe went to Israel, while the more secular Jews sought to immigrate into the United States. The price of failing to address the real problem would be decades of turmoil in both places. The state of Israel has been at war since its inception, and the tranquillity of the United States has been turned upside down by the ideological machinations of the paranoid Christians and Jews. Christian children were taught to fear the Jews, who had supported the rise of atheistic communism in Russia. They were constantly asked to pray for the poor Christians being persecuted and killed in the captive countries of the Soviet Union. They were also taught that all Jews were destined to burn in hell for having rejected and killed the Christ and that any Christian who married a Jew would suffer the same fate. Unfortunately, this insanity was not one-sided. Jews have told me that, as children, they were beaten for so much as using the name Jesus. Others were told that it was a sin to walk past a church or to listen to church bells and that Jews shouldn't associate with Christians because they might kill them, eat their flesh and drink their blood as they did to the Jew they called the Christ.[28] The father of a Jew who married a Christian would declare the child to be deceased. He would then tear his garments, recite the prayers for the dead and never speak to the child again. What madness!

Unable to address these religious fears in public, ideologues from both sides invaded the American political arena with hidden agendas. They used the euphemisms of liberal and conservative or Left and Right, but in reality these forces were driven by the same religious bigotry that has separated us all through history. The time has come to face our fears and expose the destructive nature of these divisive ideologies.

THE 1950S

When older people in America wax nostalgic, it is invariably for the 1950s. In fact, they are really longing for the period immediately following World War II. In those years, the United States was at its peak. We had emerged from the war as a military and economic giant. America was the leader of the free world, and American products were sought after as valued possessions and symbols of wealth. Optimism abounded as capitalism and Christianity ruled the land. We were the strongest country in the world, and, with troops and advisors on the ground in Europe and Asia, we set about the task of remodeling the ancient-world monarchies and fascist dictatorships along the lines of the democratic republic created by our country's founders.

There was only one problem: the communists. Though the Nazis came within a hair of defeating them, the Russians had emerged from the war as a formidable military power. Because the Church had supported the fascists in the war, many survivors (especially the Jews) were now sympathetic to the message of the communists. Secular humanists seized the moment and renewed their struggle to replace the remaining monarchies with atheistic communism. By the early 1950s, communist revolutions had overthrown the governments in China, North Korea and North Vietnam. The first in a string of hot wars between the communists on the Left and the capitalists on the Right was about to flare up in Korea, while several third world countries, like Cuba, were on the brink of being toppled by leftist insurgents. The Cold War was intense, and the propaganda knew no bounds. Fearful of the growing communist influence in America, the FBI and CIA began to infiltrate Left-leaning organizations and report their activities to the government. This was the age of anti-communist paranoia. A Senate committee held televised hearings in which Senator Joseph McCarthy tried to expose the growing communist influence in the American government. Anti-communist, right-wing Christians fueled McCarthy's investigations, and government officials, entertainers, authors, actors and businessmen who were found to have even so much as associated with a communist were blackballed in their respective professions. Others, found to be communist spies, were prosecuted and sent to jail. As these paranoid McCarthyites became more aggressive, many innocent people had their lives destroyed by what rapidly became a political witch hunt.

The media began to sympathize with the plight of these people, and by the late fifties and early sixties, the propaganda war had shifted in favor of the liberals. Now an equally destructive left-wing backlash swept the country. Soon the paranoid anti-McCarthy forces became so powerful that even the slightest disagreement with the Left could cause one to be declared a McCarthyite, anti-Semite, fascist or Nazi. Fear of being branded with one of these profane designations had a chilling effect on dialog, and it became taboo to even question a liberal doctrine, no matter how ludicrous its premise. By suppressing a discussion of the obvious flaws and contradictions in these opposing religious and political ideologies, politicians and religious leaders forced the problem underground. Consequently, what should have been a dialog became a conflict, and soon it was being played out, once again, as violence in the streets.

My Catholic parents were a perfect example of the ideological confusion of the time. The work of President Roosevelt in the Depression and World War II made them fanatically loyal to the Democratic party. My father was also a member of the Teamster's Union, and he was very sympathetic to the communist concept of labor sharing the wealth of corporations. On the other hand, their Irish Catholicism forced them to support Joe McCarthy because they were terrified of the atheism of the communists. Today we look at communists as a pretty benign force, but in those days their attempts to destabilize America were quite real, as were similar CIA operations in the Soviet Union and elsewhere.

You can readily see that fear was the driving force motivating the fanatic machinations of the ideologues on both sides. Over the years, that fear has provoked these zealots to make reckless attacks on the fundamental tenets of the opposition. One of the most destructive battles of this Left versus Right struggle is an ongoing fight to control the curriculum of America's schools. This battle, though less bloody than the hot wars, is more insidious and equally destructive. The battle lines were drawn between the proponents of Darwin's theory of evolution on the Left and those who advocated biblical Creation on the Right. The prize was political control over the ideological indoctrination of America's schoolchildren.

"GIVE ME A CHILD UNTIL HE IS SEVEN. . ."

Karl Marx was a contemporary of Darwin, and an outspoken supporter of evolution. The Darwin museum in London has, as one of its prized possessions, a laudatory personal letter to Darwin signed by Marx. It is not surprising that Marx loved Darwin. One of the core beliefs of the communist movement was that in order to take over a country, they would have to discredit its vital institutions and capture the minds of the youth. Darwin's theory of evolution allowed the communists to both destabilize the core values of the Christian democracies by discrediting biblical Creation and indoctrinate schoolchildren into the cult of atheistic science. This theory did all these things and more. It also drove a wedge between parents and children, causing students to see their parents' religious beliefs as the wishful thinking of Neanderthals. Such confusion would cause children to seek guidance from sources outside the church-family sphere of influence and open the door for the Left to feed them with radical propaganda through

Hollywood, television, the recording industry and academia.

Creation science had been the basis of the American educational system since its inception, but by the late 1950s, evolution had supplanted it as the core science curriculum of the public schools. This change in American education was the direct result of one of the greatest deceptions ever perpetrated by secular humanists with the complicity of atheistic science. It revolved around the famous Scopes trial that took place in 1925, in the State of Tennessee.

The Great Monkey Trial

As early as the 1920s, secular humanists in America were attempting to remove the Bible from the public schools by replacing Creation science with Darwin's theory of evolution. This met with stiff resistance from Christians in the South. Fundamentalists were outraged that the government was using their tax dollars to teach evolution to schoolchildren, directly contradicting the religious beliefs of their parents. In 1925 Tennessee legislator John Washington Butler sponsored the first of many laws enacted by the states to make it illegal to teach evolution in schools supported with state funds. In response, the American Civil Liberties Union (ACLU) announced that it would defend any teacher charged with violating the Butler Act.

Secular humanists decided to have a teacher violate the law in order to create a test case. A young science teacher, John Thomas Scopes, was recruited to be the legal guinea pig, and the ACLU agreed to take up his defense. Scopes used a state-approved textbook to teach a lesson on evolution in his Rhea County High School science class on April 24, 1925. A few days later he was arrested and quickly indicted, setting the stage for what became known as "The Great Monkey Trial."

The ACLU dispatched its chief attorney, Arthur Garfield Hays, to Tennessee to support the defense team, which was led by the famous Clarence Darrow, a militant agnostic and anti-fundamentalist. Darrow saw the trial as a chance to destroy Creation science and drive this intellectually weak teaching from the public schools. On the prosecution's side was Christian fundamentalist and three-time presidential candidate William Jennings Bryan, a great orator who was determined to save the schoolchildren from the "satanic" theory of Darwinian evolution.

The defining moment in the trial came when Darrow called Bryan to the stand as a hostile expert witness on the Bible. Bryan was eager to cross swords with Darrow, but this was to be his undoing. Darrow challenged him to explain the apparent contradictions in the Bible: Where did Cain get his wife? Did Bryan really believe that God had created the universe in six days? With great dramatic oratory and feigned disbelief, Darrow used these controversial Christian interpretations of Genesis to badger his famous nemesis. The personal hostility between Darrow and Bryan electrified the atmosphere, and Bryan, who seemed unprepared for this type of questioning, was made to look foolish. Eventually, he was reduced to admitting that he didn't really think that the universe had been created in six 24-hour days, which shocked his supporters. From the witness' chair, Bryan lashed out at Darrow, charging him with degrading the Holy Bible. Darrow responded by saying that it was not the Bible, but Bryan's fool ideas he was degrading—ideas, he claimed, that no enlightened Christian would believe. At that point Judge Raulston adjourned the proceedings, saving Bryan further embarrassment. The next day the Judge ruled that Bryan's testimony was irrelevant because it did not speak to the right of Tennessee to ban the teaching of evolution. Nevertheless, the damage was done. Bryan had been humiliated, and the credibility of Creation science was severely damaged.

It took only nine minutes of deliberation for the jury to convict Scopes of violating the Butler Act, but the Judge made a technical error when he imposed a fine on the defendant. Under Tennessee law, this task was supposed to be performed by the jury. On that basis, the defense had the verdict reversed on appeal. However, the appellate court made it very clear that, in reversing the lower court's conviction of Scopes, it upheld the constitutionality of the Butler Act, allowing the state to retry the case. The government, seeing the futility of the suit, decided not to order a retrial of Scopes, who was no longer teaching in the public schools.

The great tragedy of this case was that, contrary to popular thinking, it did not settle the Creation-evolution debate. This case had nothing whatsoever to do with the origins of man; it involved the state of Tennessee's right to ban the teaching of evolution in public schools. The prosecution won the case, and the appeals court upheld the constitutionality of the Butler Act. Nevertheless, due to a crafty repackaging of this trial by secular humanists, most people are under the false impression that it settled the dispute by affirming the veracity of evolution. I recently had a heated discussion with my teenage son's science teacher, who was absolutely certain that Creation science had been soundly defeated in the Scopes trial. Is it any wonder that the majority of students in the public schools are similarly misinformed?

Most parents teach their children to believe in God and Creation. When they reach the tender age of thirteen or fourteen, these adolescents enter a Junior High School science class where their teacher, without so much as a by-your-leave to parents, dashes that "myth" to the ground and replaces it with the vaunted theory of evolution. In one fell swoop, science destroys the religious beliefs of these children, debunking the core principles on which their developing sense of morals had been based. Teachers then proceed to drive an intellectual wedge between the children and their parents, whose religious beliefs are portrayed as neanderthalic. In the name of progress, scientists who would wince at the thought of disturbing the habitat of the kangaroo mouse recklessly violate the psyche of these children at the height of their most fragile adolescent years. It is shocking how destructive these ideologues can become in the heat of battle. The casualties of this war are numbered in the statistics for teenage pregnancy, drug abuse and suicide, as well as the general decline in the morals of American society.

After the Scopes trial, evolution slowly became the dominant ideology on the origins of man. Even the science teachers who professed to be Christians ignored the obvious contradiction and pretended that it was possible to reconcile the diametrically opposed billions of years of evolution with the six days of Creation. By the late 1950s, the history book we call the Bible, which has been proven to contain solid archaeological facts, had been banned from our public schools as religious fiction, while the myth of evolution, which is totally devoid of corroborating evidence, had become the nucleus of all science curricula.

The 1960s

As the 1950s came to an end, atheistic communism was expanding its influence around the world, and the Russians were using spies and communist sympathizers to destabilize the Western governments, both democratic and dictatorial, as part of a worldwide revolution. Communist ideology was well-suited to the task of radicalizing the peoples of the third-world countries in Latin America, Asia and Africa, who still suffered under the autocratic governments that had been imposed on them by the monarchies with the cooperation of the Vatican. The overthrow of Cuba by Fidel Castro in 1955 provided the communists with a toehold in the Western hemisphere, and they were quick to exploit this golden opportunity to expand the revolution. As Castro began to export Marxist ideology to his neighboring countries, the United States was forced to shore up its position in the region. The American government quickly began funneling a large amount of military and economic aid to autocratic dictators in Central and South America to prevent the much-feared "domino effect" that would later play such a big role in the Vietnam debacle. As a result, the United States was forced to compromise its most sacred principles—democracy, freedom and human rights—for the expedient goal of staving off further communist influence in the hemisphere. Brutal dictators in South and Central America imprisoned and tortured dissidents without trials, while right-wing death squads eliminated people indiscriminately. Some Catholic clerics joined the people in resisting this oppression by developing a form of Christian Marxism called liberation theology. They rightly felt that the basic principles of Marxist philosophy were far closer to the teachings of Jesus than were those of capitalism. These priests and nuns became so radical that the pope was forced to ban the clergy's involvement in politics. Many simply ignored the order, and liberation theology was soon operating in Catholic schools and churches in the heartland of the United States itself.

This was yet another convoluted mixing of ideologies for the sake of expediency. The fact that Catholic nuns and priests were willing to identify themselves with the atheistic communists provides an insight into the extent of our ideological confusion. I'm not sure we Americans fully appreciate the potential disaster inherent in the chaos of that period of our history. If a larger, more strategic country, like Mexico, had fallen to the communists or American spy planes had not detected the Russian missile sites in Cuba, the outcome of this period might have been quite different.

In the United States, the struggle was less violent but equally disastrous. Secular humanists, fundamentalist Christians, communist insurgents, civil rights leaders, radical black revolutionaries, labor unionists and activist Jewish survivors of the Holocaust were among a host of ideological operatives seeking influence over the government and academia. Left-wing ideology was slowly winning the propaganda war, and Marxism was rapidly becoming the "in philosophy" on college campuses. Academics were justifiably frustrated with America's support for the dictators in Latin America and the slow pace of change in government policy on things like civil rights and racial segregation. These professors were part of the growing atheistic, secular humanist movement that flourished when science adopted the theory of evolution. They saw themselves as progressive intellectuals trying to drag a neanderthalic Judeo-Christian America into the 20th century. Their anti-Christian stance made them a perfect target for infiltration by communist insurgents, who sought to recruit dissidents for the revolution. By championing justifiable causes, these budding Marxist professors had a powerful influence over their students. Through the schools of journalism, a generation of graduates steeped in Marxist ideology began to permeate the media. As a result, the country took a decided turn to the left, while the conservative Christians went further into retreat. The civil rights movement gave secular humanists, who were now being identified in the media as "liberals," the noble cause they needed to attract the support of the masses. These new recruits were truly altruistic people who sincerely wanted to improve the lot of the blacks. They had no way of knowing that powerful communist operatives were trying to move them to the radical Left. After the death of Dr. Martin Luther King, Jr., the cry changed from "I have a dream" to "Burn, Baby, Burn." This was a call to violent revolution, and it resulted in many cities in America being put to the torch. Sincerely innocent supporters of peaceful change now found themselves in the middle of an ugly confrontation between radical left-wing revolutionaries and the government. These radical communists also co-opted other genuine causes like the labor and women's movements, the anti-war struggle, the plight of migrant workers and gay rights, using them to foster fear of the "fascist pigs" who controlled the government.

These same radicals caused the very core of America's moral standards to come under attack. What seemed at the time to be a spontaneous liberalization of thought was, in fact, a premeditated attack on the moral values of the Western democracies. Divorce, pre-marital sex, smoking, drinking and drug use were all glamorized by the Left, while traditional values were disparaged as primitive roadblocks to happiness. This was called the "Sexual Revolution." Aided and abetted by Christianity's ludicrous doctrines on sex, the atheistic secular humanists, Rock-and-Roll musicians and Hollywood liberals attacked every aspect of the conservative American lifestyle, and at the heart of that lifestyle was the Church itself. As one who lived through this period of history, allow me to assure the reader that there was no sexual revolution. The sixties radicals didn't invent any new forms of sex. This was merely a cover for the underlying political revolution against the positions supported by the right-wing, conservative Christians. The fascist-fighting ACLU and similar secular humanist organizations on the Left used the growing opposition to the puritanical doctrines of Christianity as a springboard to challenge conservative ideology on communism, abortion, prayer in school, funding for a strong military, Christmas decorations on public property, criminal justice and the draft. All of these attempts to destabilize the system were lumped together under the fun-sounding euphemism of the Sexual Revolution. Hedonistic young Americans jumped on the bandwagon, and it became cool to be radical. The drug culture grew exponentially, with the Left encouraging people to "turn on and drop out." Women were encouraged to burn their bras; men, their draft cards. This liberal madness reached a crescendo with the Supreme Court decision on Roe v. Wade, which opened the door for the legalization of abortion on demand.

ABORTION ON DEMAND

Abortion is perhaps the most unseemly of the ideological catastrophes that are directly related to the separation of logic and morality in the ugly divorce of science and religion. This division allowed the Church to take a legitimate biblical teaching from the ancient world and, in the absence of logic, turn it into their completely irrational position against the use of birth control here in the modern world. In ancient Israel, if a man died without having produced any offspring, the law required his next of kin to go into his widow and sire children in his name. In one such incident, when O'nan laid with his deceased brother's wife, he avoided conception by spilling his seed on the ground.[29] The scripture states that O'nan's refusal to raise up children for his deceased brother displeased God. The Church misinterpreted this piece of biblical history to mean that God was against all forms of birth control under any circumstances, even masturbation. To say the least, that was a gross misreading of the original purpose of this biblical law, which, in any case, has no place here in the modern world. It was this type of ludicrous misinterpretation of Scripture that was largely responsible for the Christian Right's deranged assault on human sexuality. An indication of the hierarchy's deviant mentality can be seen in the fact that they required priests and nuns to remain celibate as a sign of their holiness, suggesting that those who engaged in the sexual pleasures of marriage were less than fully pure in the eyes of God. These frustrated clerics invaded all aspects of Christian sexuality, teaching believers that their natural hormonal responses were actually the impure desires of sinners who had fallen victim to the snares of the devil. Sex for any purpose other than procreation was forbidden; pleasurable sex, even on the marriage bed, was condemned, and masturbation or the use of a contraceptive was declared to be a mortal sin punishable by an eternity in the fires of Hell.

In an equally perverted backlash, secular humanists made it their business to convince people that the primary purpose of sex is pleasure and that sexual promiscuity is not only sanctioned by society, but quite safe if a condom is used. This led to an explosion of teenage pregnancies and sexually transmitted diseases, as well as a glut of unwanted babies. Humanists took the position that because these babies were an unfortunate byproduct of their amoral utopia, science should simply find a practical way to eliminate them before they leave the womb. In the beginning, relatively few abortions were performed, so it was somewhat easy to defend this procedure as helping a few women deal with difficult or emergency medical situations. However, as millions of mothers began to use abortion for less excusable reasons such as birth control and sex selection, it became increasingly difficult to defend the burgeoning numbers of babies being killed in this manner. What began as an attempt to humanely assist a limited number of women in unique situations soon deteriorated into a blind defense of amoral ideology that attacked any effort to address the magnitude of the problem. Even reasonable attempts to limit the scope or methods of abortion were condemned by secular humanists as some sort of secular sacrilege.

Like the right-wing scientists in Nazi Germany who excused their genocide as an admittedly repugnant means to a noble end, left-wing scientists now supplied the medical technology to exterminate these infants, justifying this infanticide by claiming that mankind would benefit from their experiments on "fetal tissue." Not unlike the cryptic "methodical selection" and "Final Solution" of the Holocaust, clinical-sounding terms like "fetus," "embryo," "termination of pregnancy" and, eventually, "partial-birth abortion" were employed to sanitize this massacre of innocent babies in the safety of the womb. These euphemisms were necessary because the scientists knew that anyone not steeped in their amoral ideology would find this mindless carnage appalling. In order to resolve the abortion dilemma, we will have to reexamine the origins of the Left-Right civil war that caused this catastrophe. Without first solving the science versus religion conflict, we will never be able to trust each other to use a balance of logic and morals to reach a principled consensus on issues like abortion.

The "Cold War" Gets Hot!

The government in South Vietnam had been put in place by the French with the support of the Catholic Church. The President, Ngo Dien Diem, had been groomed for his position by the hierarchy of the Church, so he was staunchly anti-communist. Here again the Left vs. Right conflict causes an irreconcilable division in a country, and once again it leads to war. The superpowers began to take sides, with China and the Soviet Union backing the atheistic communists in the North, and America, Australia and some Western allies backing the capitalist Christians of the South. President Lyndon B. Johnson expanded America's military role in the war to the point where domestic resistance began to threaten the stability of America itself. The Left used the growing anti-war sentiment among the country's youth to further their ideological goals, and, as these radicals took to the streets, the country began to come apart at the seams.

At this point in history, democracy and communism—the two ideological systems that had started out with the noble purpose of designing a system of government with which to overthrow the ancient monarchies—were locked in a deadly competition to overthrow each other, bringing the world to the brink of nuclear war. Lost was the original purpose of these movements, and the failure of their leaders to address the underlying religious strife kept them from understanding the demise of the ancient world, which necessitated the move to a new world order in the first place. Though much progress had been made in eliminating the anachronistic monarchies, we humans were still flailing about in what can only be described as the semi-dark ages, and we will remain so until we candidly address the fall of the gods and subsequent rise of man to rule the earth.

By the late 1970s, America was experiencing the full effect of the anti-fascist, liberal backlash that was part of the fallout from the Holocaust. Thirty years of paranoid, left-wing attempts to prevent a second Holocaust by driving any hint of fascism from the criminal justice system had produced liberal judicial decisions that required policemen to be lawyers and sent them to jail for any infraction of the ACLU's code of conduct. These anti-fascist crusaders spared no effort in earning their credentials, even going so far as to release hardened criminals who had confessed and been proven guilty in a court of law merely because the police failed to point out all of the left wing's legal escape hatches or because a cop appeared to use excessive force in making the arrest. It is one thing to discipline rouge cops; it is quite another to let one's ideological bent pervert judicial decisions. Implementing an iron fist to destroy the life of a policeman who becomes overzealous in the heat of an arrest, while hiring psychiatrists and armies of social workers to try to understand and explain the motivation of homicidal criminals, speaks for itself. I wonder how the judiciary or the ACLU would stack up under a similar close scrutiny of their professionalism. This insanity is clear evidence that Holocaust paranoia had made the religious Left more afraid of "fascist policemen" than wanton criminals. We must find a way to candidly address the erroneous concepts driving this paranoia.

Strange Bedfellows

In the late 1960s and 70s, our failure to face the real "Jewish Problem" at the end of World War II came back to haunt us in the Middle East. Islamic resistance to the state of Israel escalated to a jihad among Muslim countries, forcing the principal ideologues of the Cold War to make bizarre allies. The atheistic communists in the Soviet Union, bordered by radical Muslim neighbors, supported Islamic opposition to the Israeli state. We now had the spectacle of Islamic holy men receiving military assistance from the atheistic communists in the Soviet Union. The Jews, who had supported the communist revolution in Russia against the Christian Czar, were now aligned with the anti-communist Christian democracies of the West that had refused to take Jewish refugees before, during and after the Holocaust. Radical Islamic liberation fronts began plaguing these supporters of Israel with terrorist attacks. Oil-rich sheiks bankrolled this jihad, and the Soviets supplied the arms and military training. Fanatic Muslim leaders began trying to outdo each other at spilling blood for Allah. Planes were hijacked or blown out of the skies, airports were bombed and sprayed with gunfire, and Beirut, the financial center of the Middle East, was reduced to a smoldering ash heap. Christians, Jews and Muslims fought each other door to door in the streets of that once beautiful and prosperous city. Nothing was too extreme to be attempted; radical Muslims strapped explosives to their bodies or loaded them into a car and exploded themselves in crowded streets for the glory of Allah. Who at that time would have believed that one day, when they grew tired of killing each other, the leaders on both sides of this madness would be awarded Nobel prizes for promoting peace?

A Man of God in America Installs an Ayatollah of Allah in Teheran

During the presidency of the Born Again Christian Jimmy Carter, a fundamentalist Islamic revolution, headed by Ayatollah Khomeini, threatened to overthrow the shah of Iran. Unwilling to intervene, Carter relied on advisors who counseled him that Khomeini was a truly religious man who could be trusted to govern the country. Based on this assessment of Khomeini, Carter allowed the shah's government to fall. This decision immediately backfired as Iran's new leader declared both Carter and America to be a "Great Satan" guilty of many crimes against Islam, not the least of which was support for Israel. The Ayatollah then allowed the American Embassy in Teheran to be attacked by a mob of fanatics. They surrounded the entire compound and held the staff hostage for more than a year. The U.S. military attempted to rescue them, but the mission ended in an embarrassing debacle. Fallout from the Vietnam War had so weakened America's armed forces that they could no longer be counted on to defend American citizens from renegade regimes. In their zeal to keep America fascist-free, the Left had all but killed the golden goose.

President Ronald Reagan
and the Conservative Christian Backlash

At this point in history, an anomaly occurred. A charismatic conservative ran for president. This charming former actor reminded the rank and file Americans of their great heritage and called on them to restore the country to its past glory. In 1980 Ronald Reagan was elected in a landslide, and the Republican party won a majority in the Senate. President Reagan stopped the liberal agenda in its tracks, rebuilt the military, challenged the communists, secular humanists and the Soviet Union on all fronts and tried to revitalize capitalism in the American economy. Eight years later, when he left office, communism was in a state of decline all around the world and the American economy had experienced one of the longest periods of growth in its history. Soon after, the Soviet Union collapsed, the Berlin wall came down and the former Warsaw Pact countries, including Russia, were singing the praises of capitalism. The Cold War had ended, but liberal Jews, secular humanists and conservative Christians have yet to get that message. Here in America, their political battle continues unabated, and our schoolchildren remain expendable pawns in their ongoing ideological vendettas.

Vice President George Bush succeeded Reagan as president and utilized the revitalized American military to evict the Iraqi army from Kuwait in the Gulf War. This was one of the high points of his career, as he rallied a broad international coalition of forces to stop the aggressive behavior of Saddam Hussein. However, his failure to follow through on President Reagan's conservative economics cost him a second term, and he was defeated by Democrat Bill Clinton. Like Bush, President Clinton failed to realize the conservative shift in the country, and he began to govern from the left. This backfired in the 1994 elections, as Republicans captured both houses of Congress for the first time in forty years. Now the right-wing Christians aggressively moved to reverse decades of liberal policies and restore some conservative values to the government. While I am somewhat encouraged by the attempts to reverse the damage done by our flirtation with Marxism, the ignorance driving the agenda is as dangerous today as at any time in history.

LIFE AFTER COMMUNISM

With the fall of atheistic communism, Christianity experienced a revival in Eastern Europe. In Russia itself, the explosion of freedom caused an intellectual revolution. All of the old core tenets of communism, including evolution, came under close scrutiny. Recently, Dr. Dmitri Kouznetsov, a Russian scientist, scholar and author with three doctorates, toured America speaking on the untenable flaws in Darwin's theory. Like Dr. Denton, he provides his audiences with empirical evidence that evolution has no basis in fact and that it is a belief, not unlike a religious tenet. This should be earthshaking news. Dr. Kouznetsov came from Moscow, the capitol of atheism, speaking against Darwin, yet he was ignored by the mainstream media. Dr. Kouznetsov is a member of the Moscow Creation Science Fellowship, of which all 186 members have Ph.D.s. Unlike America, where the Bible is banned from the educational system, biblical history is now openly taught on the campus of St. Petersburg State University. It seems to me that the hierarchies of Western science and academia could learn something about intellectual freedom from the Russians.

In order to show the negative consequences of mixing the flawed theory of evolution with political movements like communism, Dr. Kouznetsov reveals the relationship between Darwin and Marx, suggesting that, in Russia, evolution wasn't just science, but a pillar of communist ideology. It's his conclusion that by rejecting Creation, we reject a Creator and eliminate the restraints of being accountable to a higher power for our actions. He explains that governmental leaders who are unfettered by moral absolutes will be prone to the kind of delusions that caused the tragic social and political consequences experienced under fascism, Nazism and communism.

Ironically, as the tyrannical nature of early Christianity resulted in the acceptance of evolution in the West, the inflexible dogma of atheistic communism in the former Soviet Union is now leading to the revival of Creation science in the East. Unfortunately, the fall of Marxism is being misinterpreted by many in Russia as an affirmation of the Christian religion. As a result, Christian cults in Russia are reaping a harvest of former believers from the defunct cults of Marxism and Darwinian science.

Now, in 1998, the great "Evil Empire" has fallen. The hated atheistic communists have lost and the "peace-loving" Christian capitalists have won, but somehow no one feels like celebrating. Eastern Europeans find themselves right back where they started from in 1914. The Vatican is once again a force to be reckoned with there, the Jews are facing increasing persecution and a religious war between Christianity and Islam has destroyed the former Yugoslavia. The communist efforts to dispel that country's religious hatreds with atheistic Marxism were a total failure. All those ancient fears had been lying dormant and were just waiting to resurface, proving the old adage that if you wait long enough, history repeats itself.

Elsewhere, the World Trade Center has been bombed, the Middle East is in a constant state of terror, religious fanatics have used poison gas to attack innocent people riding the Tokyo subway, Jim Jones and David Koresh have led their followers to the grave, the Federal Building in Oklahoma City has been destroyed and former atheistic scientists in Russia are joining Christian cults while their colleagues in the West become more militant in their defense of Darwin. In Western countries it is now virtually impossible to acquire a tenured position in academia without bending the knee to the Left's messiah of evolution. If these ideological machinations were not so destructive, they would be comical.

As we approach the 21st century, mankind is in a retrospective mood. We know that something is very wrong, and our frustration is reflected in the recent political upheavals in the Soviet Union and America. Many Americans feel as though they want to throw everything away and start over, but they lack the courage to take such a step. Instead, they look back to the "happy days" of the 1950s, when Republican Christians ruled the land and things seemed simpler. However, the solution to our problems will not be found in political nostalgia. Instead, we must have the courage to allow the marketplace of ideas to solve our scientific and religious confusion through open and unbridled debate: a debate that will reexamine the Creation-evolution conflict that has divided us into left-wing and right-wing zealots who are incapable of communicating with each other for the common good. It is time to do away with pejoratives like "racist," "anti-Semite," "fascist," "satanic," "infidel" and "heathen" that are designed to demonize the opposition. We must begin to trust each other's motives and jointly address our fears, which are based in ignorance. We should take a businesslike approach to these philosophical divisions and brainstorm our differences in the same way that a private enterprise would approach a managerial disagreement. In that atmosphere, all parties would be allowed to candidly express their concerns without being described in negative terms, and one's own logic would be the final arbiter on the veracity of a position.

COMMON GROUND

Having been born in 1938, I lived through World War II and the Korean War when Americans pulled together as one people to save Europe and Asia from the ideological cauldron that had engulfed the world. We Americans proudly planted the seeds of democracy on both continents, confident that it was a model government with which man could justly rule the earth. I recall as a child how happy my Irish immigrant parents were to live in a country where freedom enabled people from different cultures with conflicting ideologies to live together in peace.

I was in my twenties when that same ideological cauldron of Left and Right began to tear the fabric of our own country in the late 1950s. This division reached its peak during the Vietnam conflict, as left-wing radicals tried to break America's will to prosecute the war. Unhappy with the Christian Right's conservative hold on the government, the radical Left made an attempt to destabilize the political structure through an all-out assault on the country's core values. It was the equivalent of setting your house on fire because you didn't like your father's authoritarian rule.

Now, as I approach my 60th birthday, the consequences of that ideological struggle are all too apparent. In the 1950s, divorce was a last resort for childless couples, not something that parents did "for the sake of the children"; we had back alley abortions, but no one tried to pretend that the baby was a blob of protoplasm; we had illegitimate births, but they were a source of shame, not government subsidies; we had sex before marriage, but it wasn't flaunted on television; we used condoms, but not as handouts in elementary schools; people who received welfare felt guilty and grateful, not arrogant and entitled; cops were peace officers, not soldiers; our job training programs were called High Schools; teenage crime meant vandalism, not murder; we had real racism, but we didn't burn down cities, gun down innocent commuters[30] or release murderers back into the community as redress; and poverty wasn't used as an excuse for criminal behavior. Forty years later, we have lost our moral compass, and even though the Cold War is over, the deleterious effects of the radical Left's attack on America's core values remain. The Left must come to grips with the fact that their attempt to create an amoral society has been a disastrous failure, and the Christian Right will have to face the fact that their untenable code of morals has been incapable of withstanding the Left's assault.

It is clear that the divorce of science from religion after Galileo created an ideological rift that has been slowly destroying the very fabric of human civilization. We have seen how religious fanatics, who believe they can communicate with God, always find that He approves of their Inquisitions, Pogroms, Jihads and Crusades; we have also seen how the lack of a belief in a higher power led Darwinian scientists to feel justified in playing God in the Holocaust. As a result, it is now apparent that atheistic "amorality" is every bit as dangerous as religious fanaticism.

Thanks to the honesty of men like Peter Tompkins, Erich von Daniken and Dr. Michael Denton, it is now apparent that the two paradigms produced by the cults of science and religion do not fit reality. Man is not a descendant of the primates, nor is he under the protection of an Almighty God. We have tried both of these belief systems, and both have failed us. The systematic debunking of these theories is creating a crisis in both science and religion. It is time for scientists to repudiate Darwin and openly address the irrefutable evidence that man is an alien. On the other side, religious leaders must admit that the supernatural forces who placed our ancestors on the planet and interacted with them in the ancient theocracies have long since departed. Mankind is now alone on the earth, the sole ruler of the modern world. The benevolent truth inherent in that reality will solve the mystery of our origins, heal our ideological wounds and provide us with a shared orientation that will allow us to live together in peace.

Clergymen are constantly berating man for not having enough faith in God, but if we really think about it, of what meaning is it to God how man feels about Him? Man's faith in God is irrelevant; however, it is critical that we become aware of God's great faith in man. As a sign of that faith, He destroyed the supernatural forces that ruled the earth in ancient times and elevated humans to their current position as the preeminent beings of the modern world. In that act, God showed His great love for man and total confidence in us to rule the earth in a just and noble manner. In the ancient world, our ancestors were mere chattel. They could be bought and sold as slaves, forced to participate in mortal combat as entertainment for the demigods or used as human sacrifices in satanic rituals. In the modern world, man has outlawed slavery, and even animals have rights. Obviously, God's faith in man is not unfounded. Perhaps we're more noble than the demoralizing doctrines of science and religion have led us to believe. From our lowly position when the ancient world collapsed, man has risen to the point where he can walk in the heavens, like the gods before him, and is slowly reaching out to the stars. Space exploration and the technological revolution are shining examples of the positive influence of a non-ideological science. We must have faith in ourselves to use this rapidly developing genius for the benefit of all the earth's creatures and boldly proceed to the greatness that is man's destiny. It is time to put aside the degrading and divisive ideologies of both science and religion and embrace the uplifting new paradigm of the fall of the gods and subsequent rise of man to rule the earth.

Ding, Dong, the Wicked One is Dead

In the classic movie *The Wizard Of Oz*, there is a scene where little Dorothy sprinkles the Wicked Witch of the West with water, causing her to melt and disappear. When the people of Oz hear this, they all break into joyous song proclaiming their liberation from the tyrannical reign of this despicable character. Thousands of years earlier, the great King David spoke of a time when the wicked forces that ruled the ancient world would be destroyed and all of mankind would rejoice as they received dominion over the earth. His wonderful prophecy is recorded in the book of Psalm 37, verses 9–11, 35 & 36.

> For [Satan's minions] shall be cut off: but those that
> wait upon the Lord, they shall inherit the earth.
> For yet a little while, and [Satan] shall not be: yea, thou
> shalt diligently [seek] his place, and it shall not be found.
> But the meek shall inherit the earth; and shall delight
> themselves in the abundance of peace.

I have seen [Satan] in great power, and spreading
himself like a green bay tree.
Yet he passed away, and, lo, he was not: yea, I sought
him, but he could not be found.

Can there be a better explanation for the collapse of the ancient
world than the one given by David in verses nine and eleven?
Clearly, the evil forces that ruled that world were cut off and man
did inherit the earth. Is there a better description of modern
scientists as they try to understand who was responsible for
building the great pyramids and temples of the ancients than the
one given in verse ten? Do they not diligently consider who was
responsible for these supernatural creations and find no one?
These prophecies are an exact description of the fall of the gods and
the rise of man to rule the earth. There can be no disputing their
accuracy, nor their fulfillment. So why have we failed to delight
ourselves in an abundance of peace? The answer is ignorance. Our
failure to recognize the import of the fall of the gods and collapse
of the ancient world has prevented us from experiencing the peace
promised by David. Instead of celebrating their demise, we act as
though the evil forces that reigned in antiquity continue to rule the
modern world, and we seek evidence for that delusion in the "evil
behavior" of those whom we fear. One of the central themes of this
chapter has been how those fears have caused us to mistrust the
motives of our fellow man because we perceive them to be satanic.
As we have seen, once an adversary is tainted with that stigma, all
of the restraints of human compassion are abandoned for the
overriding goal of destroying such evil. Unfortunately, no single
blood bath, no matter how genocidal, can ever absolve an enemy's
"demonic" offspring; therefore, the jihad, crusade or holy war goes
on ad infinitum.

No place on earth are the results of this fear and ignorance more
obvious than in the modern state of Israel. Islamic terrorists are so
afraid of the "satanic" Israelis that they are willing to become
human bombs in order to eliminate some of these "demons" as they

ride to work on Jerusalem buses. As if that weren't bizarre enough, heads of state like Yitzhak Rabin, who suggest that we seek the peace spoken of by David, are declared to be demonic by their fellow Jews and targeted for assassination in their own country. It is a truly poignant tragedy that Rabin's murder took place on the very soil where man was liberated from the real demonic forces that tyrannically ruled the earth in the ancient world; yet two thousand years later, we are still fighting these anachronistic battles.

The atheists in empirical science contend that our belief in God is the cause of the problem; to the contrary, that belief has motivated man to some of his most noble achievements. Nor is hate the cause. It is foolishness to claim that terrorists blow up a busload of innocent people or assassinate their leaders to satisfy hatred. No, these fanatics can only feel justified in committing such atrocities if they believe that they are helping God stamp out evil. It is our religious belief in Satan that legitimizes this carnage. That belief has made genocide a justifiable means of destroying our "evil adversaries." Satan is dead! The only place he still lives is in the fears of the religious. It is time for the clerics of the world to hold a funeral for this despicable being and put an end to this extremely destructive pretense. Not only is Satan dead, but the well-meaning wizards of science and religion, who tried to calm our fears by pretending that they knew the answers, have been exposed. Like the sympathetic impostor The Great Oz, they must come out from behind their curtains of ritual and prestige and use a combination of logic and morals to help us come to grips with our new role as rulers of the earth. Only then will we be able to fully escape the dark ages and delight ourselves in the abundance of peace prophesied by King David. Let us begin!

Obelisk—Karnak Temple, Luxor

Colossi of Amonhotep III, Thebes

iv. We Can't All Be Right

In the hope of avoiding a civil war, Abraham Lincoln quoted the Bible when he said, "A house divided against itself cannot stand."[1] Yet, on this ever-shrinking planet we call home, the religions based on that book have brought us to the point where we're not merely divided, but fractured. From a mere 200 distinct and separate Christian denominations at the turn of the century, there are now well over 20,000. Add to that figure the growing number of sects in Judaism and Islam, the myriad of pagan religions and the cult of atheistic science, and we can begin to appreciate the size of the problem. I think it can be said without exaggeration that these divergent beliefs are rapidly bringing us to the brink of disaster. The Protestants versus the Catholics in Northern Ireland, the Shi'ite versus the Sunni Muslims on the Iran-Iraq border and the Jews versus the Muslims in Israel are but a few examples of the violence caused by these religious "differences of opinion."

Here in America, these religions are locked in an equally vicious ideological and political cold war. Forced to disguise their true identities by the separation of church and state statutes, they operate under the banners of right-wing conservatives and left-wing liberals. Nevertheless, it is obvious to anyone receiving mail from these organizations that the principal antagonists are the Christians and Jews.[2] Since the end of WWII, Holocaust paranoia has led the Jews to political activism. Understandably, they viewed any laws that were based on biblical morals as harbingers of the much-feared Christian state and fought tirelessly for their defeat. The result of these endeavors was the arousal of the very monster that they had set out to destroy. Now the Christian Right, in the guise of conservatism, is attempting to rescue America from the secular humanist, liberal Left and return it to its "Christian roots." The consequence of this very destructive, undeclared holy war is a divided and floundering America. It is time to drop the euphemisms and expose our true identities. Let us not hide behind the banners of liberal and conservative, pretending that we are locked in a political struggle to save the Constitution; rather, let us have the courage to admit that the warring factions that have destroyed Bosnia, Belfast and Beirut are already at work right here in America. I am sure that the differing factions in those wars had convinced themselves that they were merely trying to save their country from the extremists on the opposing side. How long will it be before those same factions in the United States become frustrated with "governmental bias" and decide to save us via a crusade or some new form of holy war? Already we see the consequences of ignoring this problem in the bombings of the Alfred P. Murrah Federal Building in Oklahoma City and New York's World Trade Center. Due to the escalating level of ideological violence, it is no longer possible to hope that we will be able to coexist by simply ignoring our differences. Men of honest heart will have to admit that the attempt to pretend that we are all right about God is a total failure. If Christians, Jews, Muslims, atheists and others are to live together in peace, we will have to openly address our religious fears and take a second look at those conflicting beliefs, because therein lies the catalyst for our divided nation.

WILL THE REAL MESSIAH PLEASE STAND UP?

At the center of this problem is the entity called Messiah. Prophecies in the Old Testament foretell the coming of a godly king to rule ancient Israel. All three biblical religions—Judaism, Islam and Christianity—claim to hold a unique revelation concerning these prophecies. Christians claim that Jesus was the Messiah and that one day he will return to the earth. Muslims accept Jesus as a great prophet, but claim that there will be no second coming because God called Mohammed to create Islam, and traditional Jews believe that the Messiah hasn't come yet.

In addition to these mainline beliefs, there are innumerable sub-denominational beliefs.[3] For instance, in 1908, Charles Taze Russell, founder of the Jehovah's Witnesses, predicted that the second coming of Messiah would bring the world to an end in the year 1914. When his prediction failed to come to pass, the Witnesses specified successive dates of 1918, 1925, 1975 and 1986. More recently, a sect of Orthodox Jews claimed that their leader, Rabbi Menachem M. Schneerson, was the Messiah. Unfortunately, he died in 1994. Unable to accept that reality, they placed a billboard sign along the approach to the George Washington Bridge. It informed

Picture by Monica Almeida, The New York Times 2/25/96.

drivers that the late rabbi would soon be resurrected from the dead to begin his reign as King Messiah and provided a toll-free number for information. All of these attempts to apply the ancient prophecies to the people or events of the modern world only serve to devalue biblical history and make the theologians look foolish.

The one thing we can say for certain about our divergent messianic beliefs is that they cannot all be right! Most theologians agree that, in their original intent, the messianic prophecies applied exclusively to the ancient Israelites. The confusion arises from the Muslim and Christian contention that, because the leaders in Jerusalem refused to accept Jesus, God was forced to change His plan and seek followers outside of the lineage of Abraham. This premise has many basic flaws. It ignores the fact that great multitudes of Jews accepted Jesus as the Messiah—so many, in fact, that the leaders in Jerusalem felt threatened. In the New Testament we find the following statement by the Apostle John.

> Then gathered the chief priests and the Pharisees[4] a
> council, and said, What do we? for this man doeth
> many miracles.
> If we let him thus alone, all will believe on him: and the
> Romans shall come and take away both our place and
> nation. (John 11:47-48)[5]

Here we see the Pharisees and chief priests concerned that if they didn't intervene, the supernatural powers evinced by Jesus would sway all the Jews to declare him Messiah. Obviously, the number of Jews who followed Jesus was substantial.

Another serious flaw in the current beliefs concerning Messiah is that God promised Abraham that the chosen people would come exclusively from his lineage.[6] To suggest that the Almighty God was moved to break His word to Abraham because some Jews in one generation rejected the Messiah is ridiculous. It portrays God as a capricious and disloyal quitter who abandons a plan at the first sign of a problem. It also implies that God made a mistake when He chose the descendants of Abraham and, consequently, was forced to admit His error and seek followers among the non-Jews. If that theory is right, then God is not infallible and the Bible is a hoax. It is absurd to assert that God made a mistake when He picked the descendants of Abraham as His chosen people. The God of Abraham does not make mistakes. It is equally ludicrous to claim that He would allow the rebellious actions of a few Jews to force Him to break the pledge He made to Abraham and invite anyone from any tribe or kingdom to become one of the "new chosen people." The Old and New Testaments are full of scriptures, like the two that follow, which state that the covenant God made with the ancient Israelites was everlasting and eternal.

> . . .I made you to go up out of Egypt, and have brought
> you unto the land which I sware unto your fathers; and
> I said, I will never break my covenant with you.
> (Judg. 2:1)

> My covenant will I not break, nor alter the thing that is
> gone out of my lips. (Ps. 89:34)

These verses make it very clear that God had promised Abraham that not only would He not break the covenant, but that He would never alter it in any way.

How, then, would God handle an incident where Abraham's descendants totally rejected Him? Such a situation occurred while Moses was on Mt. Sinai receiving the Ten Commandments. When he failed to return quickly enough, the Israelites became fearful and broke the covenant by building the Golden Calf. God's statements to Moses concerning this transgression are very revealing.

> And the Lord said unto Moses, I have seen this people,
> and, behold, it is a stiff-necked people:
> Now therefore let me alone, that my wrath may wax
> hot against them, and that I may consume them: and I
> will make of thee a great nation. (Ex. 32:9-10)

Please notice that God did not say that He would break His covenant with Abraham and send Moses to the heathens. To the contrary, God said that He would eliminate the disloyal Israelites and raise up a new nation from the loins of Moses. That solution would allow God to create a new chosen people while still keeping His promise to Abraham, because Moses was of the chosen lineage. This scripture makes it very clear that God fully intended to keep his covenant with Abraham by selecting His chosen people exclusively from the physical lineage of the ancient Israelites.

THE SOLUTION

The Old Testament declares that the Israelites would be God's chosen people as long as the sun and moon traverse the heavens. In the following quote of that scripture in Jeremiah, God also states that the New Covenant (New Testament) was specifically limited to the house of Judah and the house of Israel.

> Behold, the days come, saith the Lord, that I will make a new covenant with the house of Israel, and with the house of Judah.
> Not according to the covenant that I made with their fathers in the day that I took them by the hand to bring them out of the land of Egypt; which my covenant they brake, although I was a husband unto them, saith the Lord:
> But this shall be the covenant that I will make with the house of Israel; After those days, saith the Lord, I will put my law in their inward parts, and write it in their hearts; and will be their God, and they shall be my people.
> And they shall teach no more every man his neighbor, and every man his brother, saying, Know the Lord: for they shall all know me, from the least of them unto the greatest of them, saith the Lord: for I will forgive their iniquity, and I will remember their sin no more.

> Thus saith the Lord, which giveth the sun for a light by
> day, and the ordinances of the moon and of the stars for
> a light by night, which divideth the sea when the waves
> thereof roar; The Lord of hosts is his name:
> If those ordinances depart from before me, saith the
> Lord, then the seed of Israel also shall cease from being
> a nation before me for ever.
> Thus saith the Lord; If heaven above can be measured,
> and the foundations of the earth searched out beneath, I
> will also cast off all the seed of Israel for all that they
> have done, saith the Lord. (Jer. 31:31-37)

These verses make it crystal clear that God was going to stand by His choice of the ancient Israelites as His chosen people and that the New Covenant, or new agreement, was limited to the descendants of the houses of Judah and Israel. There is absolutely no mention anywhere in the Bible of expanding this privilege to include non-Israelites in either the ancient or modern worlds. An identical scripture in the New Testament book of Hebrews quotes this promise verbatim, totally disqualifying the Christian and Muslim claims that God later changed his mind.[7] These scriptures force us to admit that the God of Abraham was limiting the work He was doing among men to the ancient Israelites. Both the Old and New Testaments of the Bible are historical accounts of His interaction with them in the ancient world. It is just as ridiculous for modern Jews to reject the New Testament portion of that history as it is for the Christians and Muslims to claim it.

The source of this confusion among Christians concerning the chosen people are scriptures in the New Testament that speak of the disciples of Jesus being sent to the gentiles. The solution to that apparent contradiction lies in the division of the two kingdoms of ancient Israel. After the death of King Solomon, a power struggle ensued, and, in order to settle the dispute, the Israelites agreed to divide the kingdom. Ten tribes became the Northern Kingdom of Israel, while the two remaining tribes became the Southern Kingdom of Judah. The Southern Kingdom remained loyal to the rightful king and maintained control over the city of Jerusalem. In addition, they continued to circumcise their male offspring and register their lineage in the temple within eight days of birth. In short, they kept the law and thus maintained their position as the chosen people. On the other hand, the ten tribes of the Northern Kingdom abandoned the practice of circumcision, lost the records of their lineage and were later conquered and assimilated by the Assyrians. The Jews of the Southern Kingdom considered their kinsmen from the ten tribes of the Northern Kingdom to be gentiles and called them by the vulgar title *akrobustia*, or *foreskinners*, because they were uncircumcised. The Israelites of the lost tribes of the Northern Kingdom are the gentiles referred to in the New Testament. As we have seen in Jeremiah's prophecy, God had promised to restore them to the Kingdom before the end came.[8]

An objective reading of the New Testament Scriptures will reveal that Jesus and his twelve apostles were sent by God exclusively to the descendants of Abraham. This reality is affirmed by the words of the Messiah himself:

> . . .I am not sent but unto the lost sheep of the house of
> Israel. (Mat. 15:24)

It is worth noting that the Greek word translated *sent* here implies that Jesus had been commissioned by God to a specific people in the same way that we send ambassadors to other countries today. When the President of the United States sends an ambassador to Mexico, that emissary is not at liberty to alter his commission and go to Japan. He is bound to limit his ambassadorship to the people specified in his commission. In this verse, the Messiah clearly states that his commission was limited to the ancient Israelites. It was not within his purview to change God's instructions and take his message to non-Israelites.

In the beginning, the disciples went exclusively to the descendants of Israel in the Kingdom of Judah (the Jews). Later, the apostles Peter and Paul received special commissions from God to take the good news of the restoration of the Kingdom (government of God) to the gentiles of the Northern Kingdom and offer them an opportunity to be rejoined with Judah in the restored Kingdom.[9] Over the years, their commission to the gentiles has been misinterpreted to mean that because some of the Jews rejected Messiah, God sent the disciples to the heathens. We must reject this erroneous theory and come to terms with the fact that only the descendants of Abraham, through the lineage of Isaac and Jacob, could claim to be the chosen people. It is time to face the awful truth that Jesus was the prophesied Jewish Messiah, that all of his followers were Israelites and that the gentiles referred to in the writings of Peter and Paul were the descendants of Abraham through the lineage of the lost tribes of the Northern Kingdom.

WE CAN ALL BE WRONG!

The Catholic Church claims that Messiah has already come and that he is actually present in their Holy Communion. This doctrine goes on to claim that after the Communion wafer is blessed by a priest, it becomes God incarnate, and all Catholics are required to worship this otherwise ordinary piece of unleavened bread. The Jews claim that Messiah has not yet come and that they are the chosen people waiting for his arrival. The Muslims claim that because both the Christians and Jews failed His test, God sent Mohammed to start yet another chosen people, and the cult of science holds that all of these religions are based in myth, while scientists are seeking the truth through Darwinian research.

The disparate nature of these beliefs forces us to admit that we can't all be right. For instance, if the Jews are right, then the Catholics must be wrong. If that's true, we should all convert to Judaism and wait for Messiah. On the other hand, if the Catholics are right about Messiah, and God is actually present in their Holy Communion, then the Jews must be wrong. In that case, we should all convert to Catholicism and join them in bowing down in adoration before the Communion wafer. As for Islam, this religion didn't start until 600 years after the fall of Jerusalem, and the cult of atheistic science didn't adopt its theory of evolution until the 19th century. It seems that the further we get from the genesis of these events, the more confused we are becoming.

While it is obviously true that we can't all be right, we can all be wrong. It is my position that a misinterpretation of the facts early in history has skewed our perception of reality; consequently, we are all arguing from erroneous positions, making it impossible to reach agreement. We must objectively reexamine all of the historical evidence in an effort to determine where the truth resides. The solution to these problems concerning Messiah has been there all along, but due to our preconceived doctrines, it has gone unnoticed. My hypothesis declares that Jesus was the prophesied Messiah and that at his second coming in A.D.70, he destroyed the supernatural forces that ruled the earth in the ancient world, bringing an end to God's dealings with man. Accepting that reality will free us from our current ideological confusion and enable us to pragmatically address our new role as rulers of the earth.

Ramses II, Karnak

Author at ruins of Inca Temple, Machu Picchu

v. Helping The Believers Be Honest

After the fall of Jerusalem, the faithful followers of the God of Abraham, Isaac and Jacob had been removed from the earth in the Rapture. All that remained of the ancient world's theocracies were the false apostles of Messiah (founders of Christianity), the pharisaic system of the Jews (forerunners of rabbinical Judaism) and the hierarchy of the pagan religions. Over the centuries, infighting has caused individual groups of believers to separate from these mainline denominations and form literally thousands of new religions, each claiming to be the voice of God among men. If there is one statement that should leap from the pages of this book, it is that man's ability to communicate with either the biblical or pagan gods is finished. There will be no further contact between the deity and mankind in this existence.

In the Old Testament book of I Kings, chapter 18, the prophet Elijah gives us an example of how we should test the credentials of religious leaders to determine whether or not they represent the God of Abraham. At that time there were two religions operating in Israel. One worshipped the god Ba'-al; the other, the God of Abraham. In a confrontation with the king, Elijah said that it was wrong for Israel to remain torn between two religions. He suggested that the matter be settled by a contest to determine which prophet represented the one true God. King Ahab approved of the idea and called the people of Israel out to Mt. Carmel. The contest required the true prophet to have God send fire down from heaven and consume an offering. Because there were 450 prophets of Ba'-al and Elijah was alone, he allowed them to go first. They prepared their offering and began to pray and entreat their god to give them a sign, but there was no response. At about midday, Elijah began to mock their efforts. He encouraged them to chant louder and taunted them with barbs like: Perhaps Ba'-al is pursuing, or on a journey, or maybe he is sleeping?[1] This infuriated the prophets, so they cried louder while leaping onto the altar and cutting themselves with lancets until blood gushed out on them. These wild religious rituals continued at a frenzied pitch until evening, but there was no voice or answer, nor any sign from Ba'-al.

As evening fell, Elijah repaired the altar of God that had been left to deteriorate by the Israelites, using twelve stones—one for each of the twelve tribes of Israel. He set wood for the fire and placed his offering on the altar. The prophet of God then called the people near to him and gave them instructions to dig a trench around the altar. He then asked that they drench the sacrifice with four barrels of water, soaking the altar, sacrifice and wood. After this was repeated three more times, the water ran down and filled the trench.

> And it came to pass at the time of the offering of the
> evening sacrifice, that Elijah the prophet came near,
> and said, Lord God of Abraham, Isaac, and of Israel,
> let it be known this day that thou art God in Israel,
> and that I am thy servant, and that I have done all
> these things at thy word.
> Hear me, O Lord, hear me, that this people may know
> that thou art the Lord God, and that thou hast turned
> their heart back again.
> Then the fire of the Lord fell, and consumed the burnt
> sacrifice, and the wood, and the stones, and the dust,
> and licked up the water that was in the trench.
> And when all the people saw it, they fell on their faces:
> and they said, The Lord, he is the God; the Lord, he is
> the God. (I Ki. 18:36-39)

That, dear reader, is an answer to prayer. There can be no doubt that Elijah's God hears and answers when His servant calls on Him. Let those in our modern-day religions who claim that we can call on the God of Abraham first show us that they can call on Him. I suggest that we challenge the leaders of Islam, Judaism and Christianity to a contest in New York's Central Park that would be called "The Day of Elijah." There they will be given an opportunity to ask God, as Elijah did, to let it be known that modern-day clergymen can call on the God of Abraham and that, like the great prophets of old, they are authorized to act on His behalf. Let them emulate Elijah and give us a sign to verify their claim that they are God's authorized representatives and that all they are doing has His approval. Any who cannot produce supernatural evidence of their authority should have the courage and honesty to face the awful truth.

In the New Testament, Jesus gave a description of the signs that would identify his disciples.

> And these signs shall follow them that believe; In my
> name shall they cast out devils; they shall speak with
> new tongues;
> They shall take up serpents; and if they drink any
> deadly thing, it shall not hurt them; they shall lay hands
> on the sick, and they shall recover. (Mark 16:17–18)[2]

All of these signs were demonstrated by the apostles and disciples as evidence of their authority; however, when asked to produce just one sign, the standard response from Christian leaders is to quote scriptures that they claim warn us against testing God. I would like to make it clear that I do not question God's power. We all know that God is capable of performing miracles. Like Elijah, I only suggest that we test the clergy's credentials. If, as he said, it was wrong for the people of his day to be caught between two religious beliefs, how can we justify the fact that there are literally tens of thousands of denominations of Judeo-Christianity today? If the monotheistic teaching of ancient Judaism is correct, then there can only be one truth. It is axiomatic that the one God of Abraham speaks with only one tongue. Therefore, I feel that it is quite reasonable to ask the leaders of these religions to comply with the litmus tests that Jesus and Elijah gave us for the specified purpose of distinguishing the true prophets of God from the fakes.

Modern-Day Religion's Unanswered Prayers

In Biblical times, only the prophets and priests had the ability to communicate with God. Now Judeo-Christianity would have us believe that anyone can approach Him. People who would tremble at the thought of approaching someone as lowly as the governor of their state have been led to believe that they can approach the Almighty God of this universe and ask him for anything from the basic necessities to forgiveness for the most heinous crimes.

Prayer, as described by the denomination to which I belonged, was a conversation between God and man. In this conversation, one could intercede for others, make requests or "just spend time with the Lord." The answer to these prayers never comes in a voice heard by the ear; instead, God supposedly places the answer in your mind or merely grants the request by a change in the situation. This leaves one unsure as to whether or not he has received an answer. The way the Catholics go around this problem is to tell you that if you don't get an obvious answer, the answer is no. I consider communication to be a conversation between two sources of knowledge. First, one speaks, and then the other responds. To say that no answer implies a negative response is the equivalent of calling your best friend to borrow ten dollars and, when he doesn't answer his phone, accusing him of refusing you the favor. When Elijah called on God, the answer was immediate and obvious. How does that compare to God's response to the endless prayers of Judeo-Christian clerics asking for peace in Bosnia, Belfast, Beirut and Jerusalem? These public prayers have had about as much success as did the prayers of the prophets of Ba'-al. The most damaging evidence against Islam, Jusaism and Christianity is their inability to show any evidence of being able to communicate with God.

A rabbi whose fourteen-year-old son had died of an incurable illness wrote a book on the subject of prayer, titled *When Bad Things Happen To Good People*, that made the bestseller list. In it, Rabbi Harold S. Kushner told the story of how he dealt with his child's long and debilitating illness. His whole congregation, as well as the local Protestant and Catholic churches, were all praying for his son, and yet the boy died. Angry and hurt by God's failure to respond to his prayers, the rabbi went into a period of deep soul searching. Out of that experience came a truly heart-wrenching book that attempts to reexamine the Judeo-Christian paradigm in order to help other grieving believers come to terms with the failure of God to answer their prayers in desperate situations. In an attempt to resolve this dilemma, he compiled a list of the possible reasons for not getting what we pray for. Most were standard excuses like the petitioner was unworthy, the prayer lacked fervor or God knows what is best for us, but others were more facetious like "prayer is a sham, God doesn't hear our prayers."

Finding all of these unacceptable because they lead to negative conclusions, Rabbi Kushner sought a more palatable explanation—one that would allow him to continue in his work of encouraging the members of his synagogue to pray. In an effort to reconcile God's failure to respond to the urgent prayers of a devoted rabbi, he determined that God must not be all-powerful—that once the laws of physics are at work, the Almighty, being totally moral, cannot break the laws He Himself has created. It was a truly ingenious solution to an almost insurmountable problem, making it possible for Rabbi Kushner to put aside his doubts, forgive God, begin to heal the loss of his son and continue in his profession. Though I disagree with the rabbi's conclusion, I salute his efforts to address the flaws in the Judeo-Christian doctrine of prayer. My quarrel is not with men like Rabbi Kushner; it is with a religious paradigm that so tortures one of its most loyal supporters as to force him to turn logic inside out and reduce the prowess of God Himself in order to accommodate its teachings.

In the introduction to his book, Rabbi Kushner states that the death of his son forced him to rethink everything he had learned about God and His ways. At a time when it should have been an important source of peace and security, his religious belief left this rabbi tortured and groping for an answer to his dilemma. Rabbi Kushner's experience in religion is a mirror image of Dr. Denton's description of the crisis in science. In his chapter on "The Priority Of The Paradigm," Denton states that even though scientists know that there is no evidence to support the paradigm of evolution and that all the recent research is negative, they continue to find ways to prop it up rather than accept reality. Kushner's book provides 148 pages of evidence revealing the physical and psychological mayhem caused by the religious teaching that God is interacting with man. He recounts story after painful story of true life experiences exposing the negative effects of this doctrine on the lives of those he has ministered to over the years. Yet, in the end, Rabbi Kushner still tries to make the case that even though God is unable to change our circumstances, He is there, through religion, to help us cope with the consequences of our misfortunes. Like the scientists, he is forced to ignore the obvious in order to preserve the paradigm. Rabbi Kushner's facetious suggestion, "prayer is a sham," is the real reason for our failure to get answers to our prayers, but, understandably, it could not be a serious consideration for a man in Rabbi Kushner's difficult position. Nevertheless, *When Bad Things Happen To Good People* is an honest attempt to confront the deficiencies of prayer in the modern world. I encourage everyone who believes in that doctrine to read it before that belief brings them to a life-crisis like the one that befell Rabbi Kushner and the others described in his book.

In another passage, Rabbi Kushner wonders how we can avoid feeling angry, hurt or unworthy when God fails to answer our prayers. He creates a scenario where he asks the reader to imagine the mind and heart of a permanently blind or crippled child who was raised on pious stories where people prayed and were miraculously cured. He then asks the reader to imagine the same child sincerely praying that God will make him whole, only to learn that his handicap is permanent. The rabbi goes on to ask if it is difficult to understand how that child's anger would be turned against God and those who told him those stories, or inward at himself, when he realizes that he will not be cured. Kushner concludes by asking if there is a better way to teach children to hate God than to suggest that He could have cured them, but for their own good decided not to.

I ask the reader to apply that same scenario to the children living in the ghettos of this country, where there are more churches than bars. These children are taught to pray and ask God to help them escape their overwhelming state of poverty. Is it any wonder that our jails and courtrooms are filled with angry young men who have vented that anger on themselves, their families and society? I have also seen numerous documentaries on the Holocaust where pundits sit and scratch their heads in wonder as to what could have caused such anger and hatred among the Germans. These same experts claim to be at a loss to explain the anger and frustration that leads a Palestinian living in Israel to strap an explosive device to his body and detonate it in a market filled with Israeli shoppers. In light of all this religious anger, I would like to add one more possibility to the list of reasons for unanswered prayer: We don't get what we pray for because God is no longer dealing with man.

On a personal note, I recall an incident that occurred when I was at the height of my involvement in the Born Again movement. My wife was working, and I was at home taking care of our son Kevin. He was about a year old at the time, and I had just placed him in his walker to play at my feet while I cleaned up the breakfast dishes. I was working at the sink when I heard him make a strange cry. To my horror, I turned to see that Kevin had grabbed hold of the steam pipe with both hands. Frightened by the pain, he was unable to let go. I quickly pulled his hands off the pipe and ran them under cold water. Unfortunately, the furnace had just reached the peak of its output, and Kevin was severely burned. The doctor treated him but said that he was too little to be given anything for the pain. All we could do was try to make him as comfortable as possible. I called the church for prayer and tried to get him to take a nap. I will never forget kneeling next to my son's crib and begging God to relieve his pain long enough for him to go to sleep. Despite my prayers, poor Kevin screamed for hours. Adding to his discomfort was the fact that he was a thumb-sucker. When he finally did go to sleep, he would try to suck his thumb, which would only wake him up again. It was one of the most difficult times of my life. I couldn't understand how God could totally ignore the prayers of a truly Born Again Christian and the screams of his infant son. I tried to bury it, but secretly I was angry at God. Most of us go along in our religious lives praying for little needs like help with our grades and careers. When these prayers aren't answered, we forget about them because they don't really matter. However, when we get into a problem that requires an obvious intervention on the part of God and we don't get it, we're forced to either reject God or make an excuse for His failure to act. Most people choose the latter, and, like Rabbi Kushner, I was no exception. I recalled how, about a month before the accident, Kevin had grabbed the pipe when it was cold. I had chastised him, and something inside me said, "cover that pipe before he gets hurt." I decided that God had tried to alert me to the danger through this incident but that I had failed to heed the warning. In this way I was able to ease my anger against my deaf God and put things back in order. On the other hand, I recall seeing a *Masterpiece Theater* series, entitled "The Jewel in the Crown," where an English missionary on a journey to Dipripur was attacked by bandits. Her manservant was killed, and she was raped. She returned to her home and burned herself to death in a shed. Her suicide note read, in part, "There is no God on the road to Dipripur." Apparently she was praying to the same deaf God I was; however, being more courageous, she faced the truth.

A few years ago, the TV show "60 Minutes" did a major story

on a couple who had allowed their child to die of a disease which is fairly easy to treat medically. They were Christian Scientists, and they claimed that their church had told them to reject worldly medical treatment and rely on prayer to God for their baby's healing. I'm sorry to say that this type of teaching is not unusual among certain Christian sects. However, what made this case unique was the fact that the people involved were highly educated. Both the husband and the wife held master's degrees. The host, Mike Wallace, couldn't comprehend how such a well-educated couple could have been so foolish. As I listened to them try to justify their illogical behavior, I realized that their experience pinpoints the source of the problem. These highly intelligent parents had been attracted to religion intellectually by the magnificence of the Bible. It is, without a doubt, the most intricate and complex book ever written, replete with prophecies that were fulfilled with precision, incredible healings and many other miracles. Any open-minded person who makes an in-depth study of biblical history cannot help but come away with at least the suspicion, if not the conviction, that it is not the work of mere men. Once this couple became awed by the Bible, they were easy prey for people who, either through cunning or ignorance, misconstrue the historical accounts of God's interactions with man in the ancient world to be the basis for an active religion. I'm not excusing the foolish actions of this couple; I'm merely making the point that when we take the promises God made exclusively to the ancient Israelites and apply them to ourselves here in the modern world, very often the results are disastrous.

It was precisely this type of misapplication of Scripture that kept modern Jews docile in the hands of the Nazis. Instead of taking steps to escape before the onslaught, many depended on prayers to God and rabbinical promises of a messianic intervention. I have read several books written by Holocaust survivors which relate stories of how the younger Jews in the camps wanted to start uprisings against the Germans but were stymied by their rabbis, who claimed that Jews were to wait on the Lord for their redemption. The rabbis would then quote prophecies which they wrongly interpreted as indicating that their deliverance by the coming of Messiah was imminent. It is a terrible truth, but prayer to a god who didn't answer and the misinterpretation of Scripture by rabbis helped the Nazis implement their Darwinian methodical selection experiment by keeping the Jews docile in the camps. The great promise implicit in the "Never Again" slogan adopted by those who survived the Holocaust will never be achieved until we all find the courage to address the role that our erroneous scientific and religious beliefs played in that disaster.

In one Holocaust book, titled *Of Blood And Hope*, author Samuel Pisar said that since the end of the war he no longer felt threatened by anti-Semitism. He further stated that he did not fear another Holocaust because he felt that Judeo-Christianity would go under forever if such things were ever allowed to occur again. As I read his book, some questions came to my mind. Why didn't Judeo-Christianity go under after the first Holocaust? Are we so mesmerized by these religions that it would take another disaster to snap us into reality? And will the looming Judeo-Islamic holocaust qualify as a reoccurrence of this madness? We should all ask ourselves what kind of a disaster it would take to get us to look at our own beliefs with a critical eye and then decide whether or not we should wait for such a catastrophe to motivate us.

In another passage from his book, Mr. Pisar wrote of his separation from his family in a concentration camp. He stated that as he watched the Gestapo direct his mother and sister to a different line than the one that he and his father were on, a rage against God and man surged in his chest. In an uncontrollable outburst, he raised his fist to heaven and cried out against God, calling Him a monster. It seems incredible, but many of the victims of this horrible crime have vented their anger and frustration against God, while science and religion are left virtually unscathed. This irrational behavior is indicative of the depth of our confusion. To all of the survivors of the Holocaust, with sympathy for their loss and admiration for their triumph over their religious and

Darwinian torturers, I respectfully submit that the God who created man did not create the Holocaust; however, the false gods that man created, did. Those false gods of modern science and religion are monsters of our own creation, and they have wreaked havoc on us throughout history.

You may say that you don't participate in any of these belief systems; therefore, it's not your problem. My response is that you don't have to be a believer to have your airplane blown out of the sky by a zealot. The terrorist bombs that destroyed Pan Am flight 103 over Lockerbie, Scotland and the day-care center located in the Federal Building in Oklahoma City did not make distinctions. All of those innocent passengers and children went to their deaths to avenge some imagined affront to an angry believer's god. Nazi scientists didn't ask the Jews and Gypsies if they believed in Darwin before using them in methodical selection experiments, nor do current scientists ask Judeo-Christian parents what they believe before forcing their children to join the cult of Darwinian science.

Ideological fanatics have wreaked havoc on society all through history. In World War II the allies had to defeat Adolf Hitler, who believed he was completing the work of Jesus Christ, and the Nazi scientists, who believed they were completing the work of Darwin. In Italy it was Benito Mussolini, who believed he was called by God to rebuild the ancient Roman Empire, and in Japan it was Emperor Hirohito, who believed he was a god. We have all seen films on the fanaticism of Japanese pilots who were willing to crash their planes into sleeping ships for the Emperor. And can we ever forget the believers who drove a truckload of explosives into the U.S. Marine barracks in Lebanon for Allah? More recently, the United Nations had to go to war to stop Saddam Hussein, who believes that he has been called by God to recreate ancient Babylon and annihilate Israel. As a result, the world's access to the vital oil reserves of the Middle East were put at risk, thousands of men lost their lives and the countries of Kuwait and Iraq were destroyed. In a similar vein, the Jews have incurred the wrath of their neighbors by declaring territorial rights based on biblical law. They believe that they are a chosen people operating under a mandate from God to recreate ancient Israel.

On several occasions, these and other believers have brought us to the brink of nuclear war. Mankind can no longer afford to play religious chicken now that nuclear and chemical weapons are part of the game. It was recently reported that one hundred suitcase-sized atomic bombs are missing from the nuclear arsenal of the former Soviet Union. One of these bombs could take down not just the World Trade Center, but a large piece of the island of Manhattan. The terrible destruction from the fertilizer bomb that destroyed the federal building in Oklahoma City will seem paltry next to the enormous death toll resulting from a nuclear explosion in a densely populated area like Manhattan. Beyond that, the radioactive fallout from such a blast would probably leave all of New York City and most of Northern New Jersey uninhabitable for an extended period of time. New York is the financial, governmental and communications center of the world. Each one of us must ask ourselves if our current beliefs are worth the potential disasters they expose us to and carefully consider the wisdom of the great King Solomon, who said, "pride goeth before destruction, and a haughty spirit before a fall." (Prov. 16:18)

As a result of the more recent attempts by religious fanatics to bomb New York City,[3] even as I write, there is a federal disaster drill underway in downtown Manhattan designed to train emergency service personnel on how to respond to a nuclear or chemical attack. This ideological crisis will not be resolved by the FBI any more than Hitler's "Jewish Problem" was solved by the Holocaust or the United Nations' "Plight Of The Jews" was solved by the creation of the state of Israel. If we are to eliminate the ideological fanaticism that makes this otherwise inconceivable attack a looming reality, we must find the courage to do what all previous generations have failed to do: honestly address the glaring errors in our conflicting scientific and religious paradigms. It is time for the scientists and clergy alike to get honest and face the awful truth: evolution is a myth and God is no longer dealing with man.

Granite hawk—Edfu Temple, Aswan

Massive Column—Karnak Temple

VI. PARADIGM LOCK
Educated Ignorance and the Amoral Majority

MANY PEOPLE FIND IT NEARLY IMPOSSIBLE TO BELIEVE that the highly intelligent scientists and theologians we have trusted with the education of our children could be so completely wrong on the fundamental principles discussed in this book. At first, I too found it incomprehensible. But over the years I have come to see this phenomenon as a "paradigm lock" that is caused by educated ignorance. An education is like a computer program; it sets the parameters within which all future information will be processed. Computers are designed to reject input that conflicts with their programming; consequently, a flawed computer program will reject data that is factually correct. Anyone who has had the experience of confronting a computer with a faulty program can tell you that no matter how hard you try to input correct information, it simply will not compute. A similar situation occurs when humans are educated from an early age to believe erroneous information. Though they may become accredited scholars and even achieve doctorate degrees in their respective fields, if the fundamental principles of that education are flawed, they have, in effect, been educated to ignorance. Two glaring examples of this phenomenon are scientists and theologians. Having been programmed from an early age to believe their respective paradigms of "atheistic evolution" and "a personal God," neither side is capable of assimilating factual information that conflicts with these erroneous beliefs. Obvious examples of this can be seen in the scientific rejection of factual evidence that supports a supernatural intervention in the ancient world and the inability of Billy Graham to understand why "God allowed" the little children to be killed in Oklahoma City. Like high-tech computers with flawed programs, these highly educated professionals are simply incapable of processing factual information that contradicts their early indoctrinations. Nothing can be more frustrating for the scholar who feels obligated to defend his flawed educational principles or for the iconoclast who tries to reveal evidence contrary to the accredited paradigm. It is impossible to imagine a scenario that would be more detrimental to the course of human progress. The only possible solution to this very serious dilemma is to convince the leaders of these disciplines to admit that neither side has solved the fundamental mystery of man's existence on planet Earth. That done, we can reopen the debate and freely receive new information on this subject. In effect, we must reprogram ourselves by telling our internal computers that it is okay to consider information that contradicts our scientific and religious beliefs.

The following newspaper article details one of the first academic attempts to revive this debate. Offered in the form of a thesis given to the American Association for the Advancement of Science by religious philosopher Loyal D. Rue, it reveals the mind-set of the scientists and clergymen alike as they begin to grope for a new paradigm. Please read it carefully. In my twenty years of research, I have never seen a more accurate portrayal of academia's growing philosophical crisis.

Post Bulletin, Rochester, Minnesota, July 20, 1991
by George W. Cornell, AP Religion writer.

Religious philosopher Loyal D. Rue says modern culture urgently needs a "noble lie"—a myth that links the moral teachings of religion with the scientific facts of life.

He said science "has eroded the plausibility of the Judeo-Christian myths. It has got to our heads and consciousness in such a way that the traditional myths can't be swallowed."

The myths, he said, include archaic views of the universe; a presumption that humans are at the center of existence; and the stories of Jesus' resurrection and of Moses bringing God's ten commandments down from a mountain.

Dispel the myths of religion, . . . and all that is left is nihilism, which considers life and the universe meaningless.

"Nihilism is not something that can be argued away. . . ," he said. "I assume it's true. But it is ultimately destructive," a "monstrous truth."

The myths served as a framework for religious teachings that brought about man's betterment, Rue says. Without their "integration of cosmology and morality"—of cosmic facts with idealism—people will deny fixed standards and do whatever they choose, splintering society.

Or they might embrace the "totalitarian option," which relies on government to force humans to behave, he said.

Rue, 46, a professor of religion and philosophy at Luther College in Decorah, Iowa, presented his thesis at a recent symposium of the American Association for the Advancement of Science in Washington.

A church-going but skeptical Lutheran, Rue suggests that we start all over and create a new myth—a "noble lie" that squares with what is known scientifically, something that is convincing though it may not be factual.

What would that lie be? He doesn't specify. "It remains for the artists, the poets, the novelists, the musicians, the filmmakers, the tricksters and the masters of illusion to winch us toward our salvation by seducing us into an embrace with a noble lie," he told the scientific meeting.

Perhaps, he said in an interview, it is possible to

rework, transpose and rephrase the Judeo-Christian
tradition to make it plausible again.

In any case, the illusion must be "so imaginative and so
compelling that it can't be resisted," so "beautiful and
satisfying" that all would feel they would have to accept
it, he told the meeting.

"What I mean by a noble lie is one that deceives us,
tricks us, compels us beyond self-interest, beyond ego,
beyond family, nation, race. . .that will deceive us into
the view that our moral discourse must serve the
interests not only of ourselves and each other, but those
of the earth as well."

He said this lie would present a "universe that is
infused with value. And such a universe is ultimately, I
think, a great fiction. The universe just is. But a noble
lie attributes objective value to it."

". . .The great irony of our moment in history" is that
what "we have most deeply feared"—being deceived—
"is the ultimate source of our salvation from
psychological and social chaos."

He said "a good lie, a noble lie, is one that can't be
shown to be a lie by exposing it to a known truth or to
science."

"We need a kind of myth, a story, a vision of
universality, that will get us pulling together, not just as
Americans, but that will make us one, and give us
solidarity of purpose. . . . It must be a lie that inspires
us to give up selfish interests in the service of noble
ideals. . ." he said.

"Without some kind of shared orientation, we can't
cooperate and can't have a coherent society."

"Without such lies we cannot live."

Fiction writers, using absurdity, could not fabricate a more
outrageous scenario to dramatize the incredible confusion in
academia than the actual events as they are reported in this article.
How did we get to the point where a philosophy professor feels
comfortable giving a speech to an audience of scientists, proposing
that we create a "noble lie?" This is a very serious problem that we

simply cannot afford to ignore. It is critical that we try to understand the rationale that makes Professor Rue feel justified in making such a bizarre suggestion.

First let me say that I do not fault the professor for his brainstorming. Like Rabbi Kushner, he is forced to operate within the bounds of two bogus academic paradigms, thereby limiting his options; nevertheless, he identified the real problem and had the courage to address it in an open forum. The professor is absolutely right when he singles out the incompatibility of the scientific and religious paradigms as the source of our rampant immorality. Recognizing the failure of these ideologies to produce a moral society, Rue reasons that if, as science contends, evolution is a truth and Creation is a myth, then the fault must lie with the Judeo-Christian myth. Being required by academia to use that mistaken belief as a given caused him to assume that the problem could be solved by replacing the biblical paradigm with a new myth that would "jive with the scientific facts of life"—thus the noble lie.

But why a noble lie? Merely juxtaposing these two words should set off intellectual alarm bells because it creates an oxymoron. As I studied Professor Rue's book on the subject, *By The Grace Of Guile*, I began to understand the reasoning behind his unusual recommendation. Academia's acceptance of evolutionary science as a truth evokes the concept of a universe that is devoid of meaning. That scientific philosophy has produced a generation of nihilistic college students who have lost their sense of morality. Alarmed by the potential consequences of a fully amoral society, Professor Rue is suggesting that we try to create a noble lie that will help us infuse the nihilistic universe of evolutionary science with moral values. The professor is quoted as saying that "science has eroded the plausibility of the Judeo-Christian myth," making it impractical in the post-modern world. This quote reveals that his concept of a noble lie is based on academia's declaration that ancient history is merely myth. Taking that erroneous belief to its logical conclusion, Professor Rue apparently thought that if the ancients were able to establish moral societies by deceiving themselves with theological fables, then, with the help of modern tricksters, we should be able to develop a more believable lie and use it to motivate society to behave morally. It is a huge leap of logic, but one can easily see how our erroneous paradigms would lead the professor to reach such a peculiar conclusion.

Professor Rue rightfully chided the scientists for failing to produce a wholesome paradigm when he said that without the "integration of cosmology and morality," people will deny fixed standards and do whatever they choose, splintering society. Here he cuts right to the heart of the matter by identifying the separation of logic and morals as the cause of the problem. As we have seen, this was a direct result of the breakup of science and religion after Galileo. That split left us without a universal standard, so people began to view morals from a personal perspective. The results are obvious. Professor Rue is absolutely right when he states that if we continue to use our current amoral system to educate our children without giving them some kind of noble philosophy to bind them together, we will eventually be forced to rely on a totalitarian government to keep order. Well said, professor! It is not possible to articulate a more accurate description of the societal mayhem produced by the mindless ideological jihad between the cults of science and religion. Unfortunately, efforts to save these flawed paradigms by creating a hybrid philosophy based on a more believable lie will be equally disastrous. In fact, it is the collapse of our original myths of "evolution from the primates" and "a personal God" that has precipitated the current crisis.

This article exposes the dilemma in both disciplines. The perceived inviolability of the theory of evolution has intellectually neutered scientists and clergymen alike and caused them to abandon the search for an uplifting, noble truth. Trapped by a cult-like desire to protect their beliefs and intimidated by the perverted concept of tolerance that permeates all of academia, both sides now appear to be willing to consider creating yet another myth in order to keep the peace and save the status quo. Notwithstanding, absent his noble lie, I applaud Professor Rue's efforts and endorse most of his points. I agree that our current beliefs are defunct and that we are in desperate need of a new paradigm. In fact, Professor Rue came very close to solving the problem when he said that it might be possible to rework, transpose and rephrase the Judeo-Christian tradition to make it plausible again. Here I am in total agreement with the professor. Instead of discrediting the ancient texts, we should question our current interpretations of the events they chronicle. We must put aside our preconceived beliefs and reexamine that history in an effort to find a new paradigm that will, as he said, make us one and give us solidarity of purpose. Only a noble truth can achieve such a lofty goal.

The Philosophical Hybrids

I have found that most people accept the fact that the highly educated theologians who run their religions are probably wrong in some of their denominational positions; therefore, they have no problem when I criticize the clergy. However, if I challenge science, the same people will become very nervous. It seems that when the secular humanists used Darwinian science to pull the rug out from under Judeo-Christians, many tried to calm their fears by pretending that maybe God had used evolution to create us. Scientists didn't openly challenge this delusion because they saw it as a means of weaning Judeo-Christians off of the "neanderthalic myth" of Creation.[1] This patronizing hoax was academia's first attempt to create a noble lie, and it left these philosophical hybrids with one foot in each camp. Consequently, when I attack Darwin, they rush to his defense like people protecting a god of last resort. A perfect example of this is Professor Rue. He began as a Lutheran, but when the Christian paradigm was debunked by science, he became a philosophical hybrid. Now, realizing that the debunking of the religious myth will probably lead to anarchy, he attempts to create a more believable lie that will both jive with the "Darwinian facts of life" and induce a sense of morality in his students.

Another example of this type of thinking can be seen in a scenario from Rabbi Kushner's *When Bad Things Happen To Good People*. Responding to his own questions concerning why good people suffer the tragic consequences of premature death from terrible diseases like cancer, diabetes or kidney failure, Kushner admits that he has no satisfying religious answer to such questions, so he turns to science for help. Sounding more like a devotee of Darwin than the God of Abraham, he suggests that because man has evolved to the point where we can overcome the natural selection that keeps the animal kingdom from reproducing defective offspring, we are creating our own disasters. The miracles of modern medicine are saving many babies that only a few years ago would have died; consequently, these weaker members of the human family in turn produce additional weak offspring only to die prematurely, leaving their family and friends devastated by the loss. Admitting that we could prevent such tragedies by allowing sickly children to die at birth, he asks, "but who among us, on moral grounds or simple self-interest, would agree to that?"

Here, once again, we find a noted theologian repeating the perverted concepts of Darwinian science as though they were gospel truth. As we have seen, Rabbi Kushner is wisely circumspect concerning the religious paradigm, yet, like Professor Rue, he is doubtless in his support for the unfounded precepts of evolutionary science.

In answer to Rabbi Kushner's question as to who among us would be willing to scientifically engineer society in order to avoid its human complications, I will quote some excerpts from an essay by John Leo which appeared in *U.S. News and World Report*, titled "A No-Fault Holocaust."[2]

> In 20 years of college teaching, Prof. Robert Simon has never met a student who denied that the Holocaust happened. What he sees quite often, though, is worse: students who acknowledge the fact of the Holocaust but can't bring themselves to say that killing millions of

people is wrong. Simon reports that 10 to 20 percent of his students think this way. Usually they deplore what the Nazis did, but their disapproval is expressed as a matter of taste or personal preference, not moral judgment. "Of course I dislike the Nazis," one student told Simon, "but who is to say they are morally wrong?"

Overdosing on nonjudgmentalism is a growing problem in the schools. Two disturbing articles in the Chronicle of Higher Education say that some students are unwilling to oppose large moral horrors, including human sacrifice, ethnic cleansing, and slavery, because they think that no one has the right to criticize the moral views of another group or culture.

One of the articles is by Simon, who teaches philosophy at Hamilton College in Clinton, N.Y. The other is by Kay Haugaard, a freelance writer who teaches creative writing at Pasadena City College in California.

Haugaard writes that her current students have a lot of trouble expressing any moral reservations or objections about human sacrifice.

In the new multicultural canon, human sacrifice is hard to condemn, because the Aztecs practiced it. In fact, however, this nonjudgmental stance is not held consistently. Japanese whaling and the genital cutting of girls in Africa are criticized all the time by white multiculturalists. Christina Hoff Sommers, author and professor of philosophy at Clark University in Massachusetts, says that students who can't bring themselves to condemn the Holocaust will often say flatly that treating humans as superior to dogs and rodents is immoral.

Sommers points beyond multiculturalism to a general problem of so many students coming to college "dogmatically committed to a moral relativism that offers them no grounds to think" about cheating, stealing, and other moral issues. Simon calls this "absolutophobia"—the unwillingness to say that some behavior is just plain wrong. Many trends feed this fashionable phobia. Postmodern theory on campuses denies the existence of any objective truth: All we can have are clashing perspectives, not true moral knowledge. The pop-therapeutic culture has pushed nonjudgmentalism very hard. Intellectual laziness and the simple fear of unpleasantness are also factors. By saying that one opinion or moral stance is as good as another, we can draw attention to our own tolerance, avoid antagonizing others, and get on with our careers. But the wheel is turning now, and "values clarification" is giving way to "character education," and the paralyzing fear of indoctrinating children is gradually fading. The search is on for a teachable consensus rooted in simple decency and respect. As a spur to shaping it, we might discuss a culture so morally confused that students are showing up at colleges reluctant to say anything negative about mass slaughter.

The secular humanist assault on America's Christian morals has left our campuses filled with amoral hybrids who come from Judeo-Christian homes but worship Darwin and see nothing wrong with human sacrifice, ethnic cleansing or mass slaughter. The original noble lie that created these hybrids, combined with the "tolerant" absolutophobia of academia, has so decimated the human spirit in these young men and women that they no longer feel confident to question even genocidal behavior in others. Proving the old adage that we become that which we hate, amoral secular humanists have managed to cripple man's most noble qualities while fostering the worst characteristics of the ancients in the name of tolerance. To those who would like to pretend that it is possible to keep one foot in religion while embracing Darwin I unequivocally state that God did not use evolution to create man and that, by allowing this pretense to go unchallenged, academia has proven itself to be a devious, immoral and atheistic cult that is far more dangerous than the dictatorial Church of the Middle Ages. For their own good, and that of their fellow man, let us hope that the collapse of the evolutionary paradigm will humble the hierarchy in science and provide a rude awakening for the disciples of Darwin in theology.

It is very clear from the John Leo article that the division of logic and morals between the cults of science and religion has left us without a moral compass. His statement that "postmodern theory on campuses denies the existence of any objective truth: All we can have are clashing perspectives, not true moral knowledge," cuts to the heart of the problem. Academia's ideological jihad has poisoned the minds of our children, leaving college graduates with an intellectual disdain for the concept of a universal truth and a conscience-numbing sense of tolerance. How can we ever hope to discover the truth concerning our existence if academia holds the very idea in contempt? We must find a way to restore a sense of morals in academia, help students to understand that sound judgment and measured condemnation are critical aspects of a mature human being, and encourage a scholarly search for the truth concerning the origins and future of mankind.

In his last two sentences, Mr. Leo provides some hope that the amoral quagmire of academia is beginning to change. More and more professors, clergymen, philosophers and thinkers like Loyal D. Rue, Harold S. Kushner, Robert Simon, Kay Haugaard, Christina Hoff Sommers, Erich von Daniken, Peter Tompkins and Michael Denton are finding the courage to speak out and expose themselves to the wrath of their colleagues by addressing the obvious flaws in the ideological dogmas of science and religion. These men and women are on the front line of our efforts to come to grips with the disintegration of our current scientific and religious beliefs. They all deserve our respect and gratitude for their courageous attempts to help us see the need for a new paradigm. Before proceeding, I think it will be of value to reexamine their suggestions.

Professor Loyal D. Rue: Declares the Judeo-Christian paradigm to be defunct. He suggests that we reexamine those beliefs in the hope of making them fit reality. —Amen.

Rabbi Harold S. Kushner: Declares some of the Judeo-Christian paradigm to be flawed. He suggests that we reevaluate our concept of man's relationship with God. —Amen.

Dr. Michael Denton: Provides a wealth of evidence that Darwinian science is a sham and suggests that we seek a new paradigm. —Amen.

Peter Tompkins: Provides indisputable evidence that the ancient civilizations utilized highly sophisticated knowledge in all fields of science and reveals how petty academic jealousies are preventing an honest appraisal of the ancients. —Amen.

Erich von Daniken: Questions the paradigms of both science and religion. He suggests that we reexamine the supernatural aspects of the ancients. —Amen.

John Leo: Quotes Professors Robert Simon, Kay Haugaard and Christina Hoff Sommers in order to expose the paradigm crisis in academia. He suggests that we seek a teachable consensus rooted in simple decency and respect. —Amen.

All of these exposés and suggestions have been of great value in my work, and each of these courageous thinkers, in their own way, has come within a hair's breadth of the truth. I thank and salute them all.

The Amoral Majority

Both scientists and theologians have been preprogrammed to believe in the purity of their respective paradigms; consequently, they are quite incapable of perceiving the role that their erroneous beliefs play in our moral decline. They are so confident in the virtue of their goals that they have convinced their followers that it would be impossible to have a successful society without their guidance. In the case of religious believers, I have seen this confusion even among people who agree with my position on man's inability to communicate with God. Though they intellectually accept that reality, they ask, "if we stop attending religious services, what will we use as a source of moral guidance for our children?" My answer is that I do not advocate simply abandoning these religious and social bonds, because their roots go too deep to be summarily torn from their psychological moorings. I do, however, encourage an open and honest debate within these religions. In addition, for those who choose to leave the system as well as those who choose to remain, it is essential that we honestly address the failure of our scientific and religious philosophies to produce a moral society. Western countries are in a state of moral decay, and nowhere is this more evident than in the United States. The skyrocketing rates of crime, drug and alcohol abuse, single mothers, teenage pregnancies, infanticide, teenage suicides, sexually transmitted diseases and divorce are weakening America's position as a world leader. Already some scholars and clergymen are comparing our current malaise to the fall of ancient civilizations like Rome and Greece. Here again, academia's failure to understand the true cause of the demise of the ancient world leads them to make unfounded, draconian analogies between ancient and modern civilization. As a result, instead of addressing the real problem—-the failure of our philosophical paradigms—they blame the evils of humanity and produce ivory-tower prophecies of our impending demise. It never occurs to these experts to look inward; consequently, scientists and theologians are constantly seeking external scapegoats for their failure to produce a moral society.

Religious leaders blame Hollywood, the left-wing media, secular humanist ideology and a lack of prayer in schools for our immorality. When they can't get Hollywood to carry their water, they turn to the government, proposing new ways to limit our Constitutional rights as a solution to the problem. The pro-life clergy points to the precipitous rise in the number of abortions as an indication of the magnitude of our problems and blames government for "legalizing infanticide." What they fail to mention is that the vast majority of the people receiving these abortions have been raised in Judeo-Christian homes. Obviously, their religious indoctrination has failed to induce a sense of morality that would, on the one hand, keep them from producing an unwanted pregnancy and, on the other, prevent them from having an abortion. Rather than blame their own failed dogma, the theologians attack the politicians and insist that the government pass laws to prevent their wayward disciples from murdering their babies.

Science, on the other hand, blames the "neanderthalic" absolutism and repressive sex education of the religious. In fact, the absolutophobia of academia will often produce an overreaction that only exacerbates the situation. A prime example of this secular zeal is Planned Parenthood. Founded in a backlash against Christianity's insane doctrines against the use of birth control devices, their original purpose was to provide a means for parents to plan the size of their families. From that very logical goal, they have degenerated to the point where they're promoting the issuance of contraceptives to children in elementary schools, rejecting any suggestion that schoolchildren be taught to abstain from sexual activity as a return to the outdated morals of a bygone era. It is both illogical and immoral for religion to deny parents the right to use birth control. On the other hand, it is both immoral and illogical for science to recommend that we give those devices to children in grade school.

It is no coincidence that, right before his reelection, President Clinton felt the need to veto a bill banning partial-birth abortions. I watched the President's press conference, where he was joined by a few women who have experienced this procedure. As he spoke, he seemed to be struggling, with tortured logic, to explain what he obviously felt was a difficult decision—finally justifying his veto by quoting statistics indicating the rarity of this type of abortion. It was later discovered that scientists had deliberately falsified those statistics in order to win Clinton's veto and remained silent as he quoted their phony figures at his press conference. Did they perpetrate this fraud to "protect the health of America's mothers" or to insure the supply of fetal cranial tissue for genetic research? As their atheistic paradigm keeps scientists from seeing the supernatural nature of the ancient artifacts, their amoral beliefs blind them to the egregious nature of this infanticide which they have euphemized as a partial-birth abortion. I am not unaware of the need for scientific research, but at what point do we decide that an ideology is out of control, and who will make that decision? Dare we trust the management of genetic engineering and cloning to the amoral cult that gave us the myth of evolution, the Darwinian Holocaust and the atomic bomb? Is it not the height of hubris for atheistic men who have declared their work to be "amoral" (not subject to or concerned with moral or ethical distinctions) to expect us to give them unlimited power to tinker with the fundamental laws of nature? Aren't these young scientists products of the same amoral educational system which produced the college graduates that John Leo described as being incapable of finding fault with the Holocaust? This is serious business, folks. It is absolutely urgent that scientists and theologians find a way to restore the blend of logic and morality that was lost in the ugly divorce of science and religion after Galileo. That will not be easy, because the vicious nature of this ideological clash has caused us to mistrust the motives of the extremists on both sides.

IMMORALITY AND THE FAMILY

The religious Right has recently singled out the breakdown of the traditional family as the cause of our immorality. I agree that the demise of the family is close to the heart of the problem; it is therefore imperative that we ascertain the exact cause of its decline. I blame the ideological machinations of science and religion. Their negative, contradictory and degrading doctrines have left our children floundering on a sea of relativism without a moral compass.

Judeo-Christian parents begin by indoctrinating their children into the religious paradigm of a personal God. From birth to age thirteen, the children believe this teaching and do all they can to please God. Then, just as they reach their most vulnerable teenage years, academia decides to crush all those beliefs and replace them with the theory of evolution. Even if evolution were a reality, this would be a terribly destructive thing to do to the psyche of these developing young minds. The most unsophisticated person knows that the adolescent years are a time of great stress for children. Yet scientists choose this particular time to force teenagers to decide between their parent's religious beliefs and science. The results of this madness are manifest in the dementia of our children. Their rates of teenage pregnancy, suicide and drug abuse are a testimony to the fanaticism that blinds these intellectuals to the destructive consequences of their amoral ideology.

There are two basic questions confronting all men: "Where did we come from?" and "Where do we go after death?" In response, science bolsters our self-esteem by suggesting that we are nothing more than "slowly evolving apes," while religion counters that we are "evil-natured sinners," destined to spend an eternity in the fires of hell. Is it any wonder that our teenagers, who are caught in this ideological crossfire, suffer from low self-esteem and have become ingenious at finding new areas of the body through which they can inject a mind-altering drug? If we want to reverse the precipitous decline in Western morality, we will have to eliminate these ego-destroying, negative doctrines and replace them with positive, uplifting teachings that nurture self-esteem. Scientists and theologians will have to seek common ground and find less toxic answers to the salient questions of life and death.

In the West, due to the doctrines of Judeo-Christianity, we tend to look at people in terms of good and evil. The devil is responsible for all the bad, while God, through our religions, creates the good. All you need to do is go to church where you will be taught how to tune out the suggestions of the devil and tune in to God. You are neither in control nor responsible for your actions; you're merely a pawn in the hands of outside forces. In the 1970s, Flip Wilson was the star of a very successful TV comedy show. In one of the skits, he would dress up as a woman called Geraldine. Geraldine would do mean and nasty things, then quickly wipe them away by saying, "The devil made me do it." It was a very funny skit, but a sad commentary on Western civilization, where no one is responsible for their actions—it's all the devil's fault. When a child "goes bad," neither his indoctrination nor his parents are to blame; it is merely a case of a good child being led astray by the temptations of Satan. Such was the case when John Hinkley shot President Reagan. I recall how his father and the newscasters marveled that a boy who was brought up in a good Christian home could have done such a terrible deed. John Salvi, who killed two people in a Boston abortion clinic, is another good example. Though his parents rightly blamed themselves, it is obvious that his Catholic indoctrination played a major role in those murders.

Judeo-Christian leaders often blame our moral decline on a lack of faith. Quite to the contrary, the immorality of our society is the direct result of our strong belief in a God who is willing to forgive repentant sinners over and over again—no matter how heinous the offense. I don't know if the reader is familiar with confession, but this doctrine of the Catholic Church supposedly helps one to blot out the sins that God has written down in the "big book." My wife calls it license to sin. It allows a Catholic to confess immoral behavior to a priest. The priest then gives a penance, based on the severity of the offense, usually consisting of some prayers to say. The penitent person then recites a prayer of contrition, and the priest gives absolution. Similar doctrines on forgiveness for immorality can be found in the Mourner's Bench of the Protestant Church and in Judaism's Yom Kippur (The Day of Atonement). The Jewish tradition allows one to transfer his immorality to a stone and throw it into a river. The teaching of the Evangelical Protestants is even more radical. It is based on the theory that because Jesus died for our sins, they're already forgiven, and one need only claim that forgiveness. Can you imagine the chaotic state my home would be in if I guaranteed my children total forgiveness, in advance, for any infraction of the rules of the house? These bizarre systems of forgiveness have done more to create an immoral society than all the drug pushers that have ever lived.

I recall an article in my local Catholic newspaper, *The Tablet*, in which the writer stated that his Jesuit ethics teacher had drilled him on the differences between morally right, morally wrong and morally neutral. As his teacher had explained it, if you thought card-playing was wrong and still played cards (which he assured them was an innocent pastime), then you thought yourself a sinner. If in your own mind you were already a sinner (though in fact you had not really done anything wrong), it was then easier to give in to real temptations: debauchery, cheating, stealing or worse. I totally agree with the good Father, and I think that the morally decadent Christian societies are a testimony to the accuracy of his teaching. I only hope that one day he will correlate it with the Church's doctrine of the "evil nature of man." This doctrine teaches Christian children that since the fall of Adam, man's very nature is evil. Teaching children that they have an evil nature is the equivalent of telling them from an early age that they are dumb. In either case, these doctrines will become self-fulfilling prophecies, resulting in a dumb or immoral adult. Is it any wonder that we can't build jails fast enough to accommodate the exploding criminal population?

Even among the more successful members of society, we hear of the negative results of these religious concepts. Many years ago, in an interview on "The David Suskind Show," television star Phil Donahue was asked by Mr. Suskind how he felt about his success. He responded, "I'm becoming very comfortable with myself. I'm finally beginning to realize that I'm a much better person than my Catholic education had led me to believe." I interpret his statement to mean that Mr. Donahue, not unlike millions of other Catholics, has had a hard time recovering from the assault on his self-esteem which is part and parcel of a Catholic education.

Judeo-Christianity is in the business of loading man down with guilt by judging him according to the law that God gave to the ancient Israelites and declaring him a sinner. That law didn't even apply to the non-Israelites of the ancient world, yet these religions have convinced millions of innocent people here in the modern world that they have broken "God's Law." No man today is under the law that God gave to ancient Israel; therefore, no man today can be a sinner.[3]

Several years ago, I read a book written by Emmett McLoughlin, titled *Crime and Immorality in the Catholic Church*. It compiled the statistics for the immorality of the Catholic people. His research was exhaustive, and his statistics were very disturbing. An early Catholic indoctrination had brainwashed me to believe that Catholics were more moral than other people. My mother told me to seek out fellow Catholics in the marketplace because they could be trusted. Mr. McLoughlin's statistics don't bear that out. Percentage-wise in 1962, the number of Catholics in jail in America was more than double the Catholic population of the country. His book is loaded with equally incredible statistics on mental illness and juvenile crime. The most incriminating statistics of the book pertained to Rome itself—the very seat of the Holy See. It has one of the highest abortion and murder rates of anyplace in the world.[4] I might add that Catholic Italy has also spawned some of our most sinister criminal organizations. These revelations are dramatic, but please keep in mind that Mr. McLoughlin is a former Franciscan priest who converted to Protestantism and, as such, only compiled the statistics for Catholics. When the other Judeo-Christian denominations are added into the calculations, the figures become astronomical. We are no longer able to walk the streets of this country after dark because of the immorality produced by the misguided doctrines of these religious systems. Mr. McLoughlin also quotes some statistics produced by the Catholic Church itself. In 1926 the Franciscan Fathers did a study. The census of that year showed that only forty percent of the people in America claimed to be affiliated with a religion, while in the jails that same year, ninety percent of the inmates claimed the same distinction. That meant that the forty percent of Americans who were affiliated with a religion were supplying ninety percent of the prisoners, while the sixty percent who didn't attend church were supplying only ten percent.[5] At the end of the report, the Franciscans themselves ask the obvious question: "What use religion?"

On October 11, 1984, *The Tablet* ran an article, titled "Drugs In The Schools," which compared the abuse of drugs among Catholic school students to that of the public school students. For the most part, the statistics for hard and soft drugs were identical with the exception of alcohol abuse, in which case the figures for Catholic schoolchildren were much greater than those of their counterparts in the public schools. Despite the fact that roughly sixty percent of the children in both systems were experimenting with drugs, the article took great pride in the fact that the Catholics fared no worse than the public schools. As telling as the statistics were, they failed to take into account the percentage of Judeo-Christian students in the public school system. Only when that is done can we begin to appreciate the magnitude of the problem. These shocking figures very quickly become part of the statistics for teenage muggers, drunk drivers and armed robbers.[6] We must find the courage to admit that the vast majority of the people incarcerated in our prisons are the products of Judeo-Christian homes and schools.

In that same vein, a few years ago I received a newsletter from our Republican Senator, Alfonse D'Amato of New York. It read:

> Dear Fellow New Yorker:
>
> Two-thirds of American youth try drugs before finishing High School!
>
> One of the greatest problems confronting society and our children today is the drug epidemic. Our youth have the highest level of drug use and dependence of any nation in the industrialized world. Kids are smoking marijuana, taking pills and getting drunk as early as the fifth grade! One and a half million young people between 12 and 17 have used cocaine, one of the most destructive drugs of all. I advocate a comprehensive and coordinated campaign to break the hold that drugs have on our society.
>
> To prevent destruction of innocent lives, I cosponsored

the Juvenile Justice Act. This legislation involves the
Justice Department in all aspects of delinquency
prevention.

Also, two of my anti-drug amendments recently
became law. We are doubling the number of elite
antismuggling teams in New York by adding 100 new
Customs agents and giving New York its first air and
sea squadrons to stop drug smugglers. These squadrons
will use new P-3 surveillance planes, Blackhawk
helicopters, high-speed intercept planes and boats, and
a state-of-the-art radar ship. This massive effort will
reduce the quantity of drugs being peddled brazenly on
the streets and in the schoolyards.

Is it not time that we asked ourselves why we need an armada
of navy ships and planes patrolling our coastline to prevent our
children from getting the drugs they obviously need in order to
survive in the culture we have created? Isn't it time we explored the
reasons why other cultures, such as China and Japan, have next to
no juvenile delinquency or teenage drug problems? We Americans
simply can't afford to continue to ignore this glaring defect in our
society. We must stop deceiving ourselves by pretending that some
invading army of criminals is forcing drugs on these kids. It's time
to face the truth: The mental instability of our children has created
a market for illicit drugs that even the best efforts of the criminal
world have been incapable of supplying. We must stop using the
underworld as a scapegoat and place the blame where it belongs,
at the door of those who have been given the task of guiding
the children of Western civilization into morally stable adults—
parents, academia and the clergy.

When confronted with their failure to produce a moral society,
religious leaders attempt to avoid culpability by accusing the
sinners. This is the old "blame the victim" ruse. A classic example
of this is found in another article from my local Catholic paper,
titled "Pope Travels to Naples to Denounce Crime, Corruption."[7]
This article reported how the pope went on a barnstorming-type
speaking tour of Naples, Italy, known as the city of incurable ills.

The Neapolitans received a papal tongue-lashing because of their high concentration of organized crime, drug use and political corruption, all of which had resulted in an economic decline. The Holy Father went on to give suggestions as to how the politicians should deal with these problems but never mentioned the role of the Church. Isn't Italy a Catholic country? Are not all of these Italian politicians, organized criminals, drug users etc. the products of Catholic homes and education? Isn't the Church in any way responsible for the immorality of the Catholic people? If, after 1,700 years, the Catholic Church has failed to produce a moral society in Naples and Rome, then maybe it is time for the Italian people to challenge its teachings.

Many will argue that the breakdown of morality is directly related to the rate of poverty. I agree that poverty is connected to immorality, but it has been my observation that the reverse is more common—that immorality breeds poverty. Many poor children in ghettos, instead of using the great resources of this country to move up the socio-economic ladder, are praying to God and asking Him for help. When God fails to answer their prayers, the resultant anger and lack of self-esteem, combined with guaranteed forgiveness of sins, will lead them to a life of welfare, crime and broken homes. The myth that our immorality is caused by poverty also ignores the fact that crime isn't limited to the ghetto. An article that appeared in the business section of the *New York Daily News* on March 25, 1986 quoted the U.S. Chamber of Commerce as estimating the annual loss to American businesses due to employee theft at $40 billion. The loss of productivity due to alcohol and drug abuse was put at another $40 billion. The article went on to say that if these staggering figures aren't reduced, American business cannot survive.

In the midst of all this, the clergy remains aloof, blaming the politicians for spending money on arms that could be used to help the poor. Never mentioned is the fact that most of our modern-day political divisions are rooted in their confused religious ideologies. The two great worldwide wars, the Korean War, the Vietnam War, the Gulf War, the revolutions in Central and South America, the

Israeli Wars and the Cold War between the atheistic Soviet Union and the Judeo-Christian West are but a few of the conflicts that are rooted in the ideological disputes that resulted from the clergy's failure to address the implications of the fall of Jerusalem in A.D.70. Multiplied millions of lives have been lost, and the damage to the world economy is impossible to calculate. The interest alone on the debts caused by these holy wars would go a long way toward helping the poor. If we ever find the courage to face the real enemy of mankind (the erroneous and divisive doctrines of science and religion) we will be able to destroy our weapons of war and direct our energies toward the task of building a safe world in which no one need go hungry.

The breakdown in morality is not limited to Christianity. For many years Jewish leaders could proudly point to statistics which showed them to be among the most moral people in the world, but the situation is rapidly changing. More recent statistics reveal a rising rate of drug and alcohol abuse among Jewish children. I would also point out the insider trading mess on Wall Street a few years ago and the infamous scandals involving Jewish politicians in New York City. A glaring example of this was the case of Donald Manes. He was a New York City Borough President who committed suicide in front of his wife and daughter, rather than face criminal prosecution for graft. In a eulogy after the funeral service, our then Mayor, Ed Koch, said that when Donny's works were weighed in the balances, the good that he had done would far outweigh the bad; therefore, he was certain that Donny would receive a place in heaven. The message to the children of our city? It is all right to betray your oath of office, rob the public treasury and stab yourself to death in front of your teenage daughter as long as you have done enough good to secure your place in heaven. Is it possible to concoct a more destructive philosophy?

The liberal Jewish influences in education, constitutional law, Hollywood and the politics of this country are directly related to what happened in the Holocaust. The Jews are so afraid of Christian fanatics that they feel compelled to fight them at every turn. As a result, like the leaders of Planned Parenthood, they tend to go to extremes. Often this leaves them in the absurd position of having to support immoral activities prohibited by their own Scriptures merely because they're declared "sinful" by the Christian Right. Liberal Jews seem to feel that it is necessary to accept every kind of aberrant behavior and belief in order to avoid having their own philosophy called into question. Is this open-minded tolerance, or just a disguised attempt to shield the Jewish paradigm from scrutiny? It is not progressive to condone immorality, and being defensive is totally antithetical to the very essence of liberalism. The sign of a true liberal is one who is willing to have his personal beliefs challenged in the court of public opinion. At any rate, it is clearly counterproductive to an open society to allow the paranoia generated by Nazi violence to stifle a healthy criticism of the Jewish paradigm.

I have known both very liberal conservatives (people willing to have their right-wing politics challenged in an open forum) and radically conservative liberals (people who think that their left-wing politics are above question). In fact, I had always considered myself to be a liberal until the Left perverted its meaning. I then changed my self-description to that of a realist, but kept liberalism's most noble credos of self-examination and an openness to external scrutiny. As such, I stand ready for your constructive criticism. However, I do not consider childish name calling like "Catholic basher," "anti-Islamic," "Neanderthal," "hater" or "anti-Semite" to be constructive. To me, these epithets are merely paranoid attempts to avoid an external critique of one's beliefs.

Religious paranoia has made it impossible for the government to deal sensibly with constitutional law on issues such as the separation of church and state, pornography, abortion, criminal justice and the death penalty. The virulent self-righteousness coming from both sides forces Christians and Jews to equate the art of political compromise to making a deal with the devil. The fact that the conservatives tend to base their political positions on their interpretation of the Bible is perceived by liberal Jews as a continuing effort to create a Christian state. This type of fear goes back to the very founding of the country. Due to the fact that throughout history religious theocracies have committed terrible atrocities, the founders tried to place a constitutional wall between church and state. The Jews themselves have used those precepts to prevent us from including Jewish history (the Bible) as a part of the curriculum in our schools. Meanwhile, the study of pagan history, including Homer and Virgil, is considered essential to a well-rounded education, and the myth of evolution is taught as scientific fact. It is a sad truth that while the separation of church and state issue has been manipulated to keep our children ignorant of biblical history, government officials from the White House to the State Houses are courting religious leaders and allowing them to influence policy decisions at every level of government. It's time to expose the shell game that distracts us by banning biblical history from our schools while politicians are climbing into bed with the clergy "Left and Right." We must put the Bible back in the schools to be read and challenged like any other history and use the separation clause as the founders intended, to drive religious influence from the government.

THE MARGINALIZATION OF WESTERN FATHERS

The cornerstone of a sound family is the father, and it is my opinion that the absence of a strong father in Western culture is directly responsible for our current malaise. In Eastern cultures, where the father is still a strong authority figure, the traditional family remains intact; consequently, it might be wise to look at the East-West differences for clues to the cause of our problem. One of the key characteristics of the Eastern father is that he is the moral leader of the family. In Asian cultures, the father is the repository of ancient wisdom and knowledge that guides his children to peace. This wisdom gives him an honored position in society, which enhances his ability to rear children who will understand and respect authority. Juvenile delinquency is practically unheard of in

Japan and China. In Eastern societies, the father is expected to shape his children into morally sound adults; any infraction of the law by his children is therefore considered to be a sign of weakness in him. When a child does something wrong, he not only brings disgrace on himself but on his father and the whole family as well. This helps to maintain civil standards and gives those societies a sense of order.

Obviously, at one time our culture had similar moral standards, but something changed. The question is, What caused Western parents to lose their moral authority in the home? A major portion of the blame can be laid at the door of our religions. In the West, the Church was a primary source of education for the poor; therefore, they had a disproportionate influence over family life. I recall that my Irish Catholic father was more intimidated by a priest, who he believed represented Almighty God, than a policeman, who represented New York City. When a priest or nun approached him, my father would take off his hat, bow his head and wait to be addressed. Even the titles adopted by the Church hierarchy (Holy Father, Mother Superior, Sisters and Brothers) were subtle infringements on the integrity of the family. To this day, the clergy is considered to be the singular repository of wisdom and morality in the West. Their convoluted code of ethics based on sin and forgiveness has alienated most modern fathers, causing them to abandon the duty to shape the character of their children. This macho disdain for the trappings of religion has, by default, given the Church as much control over their families as did my father's humble submission. Now, when a child asks his father a question pertaining to ethics or God, he will refer him to the minister, priest or rabbi. When the child goes to these men, in some cases he is taught to pray to the Eucharist for guidance, or he may be told of a God who is preparing to burn him in the eternal fires of hell for an infraction of religious law. Even though most parents have long since ceased to believe these tenets, they insist that their children attend religious services in order to give them some moral guidance. A young woman who works with my wife describes herself as an agnostic, while her husband claims that he is an atheist. Yet to pacify their extended families (and perhaps some unacknowledged fears of their own), they send their two children to Sunday school for religious instruction. Teenagers resent being forced to participate in these rituals merely to satisfy parental guilt, and as soon as they're old enough to make their own decisions, they abandon the practice. The bizarre rituals of Judeo-Christianity give children the feeling that their parents have lost touch with reality and can no longer relate to the problems of the real world. This loss of respect influences other decisions, and teenagers from religious families tend to avoid sharing their secular problems with their parents. As a result, they turn to their peers, Hollywood, the recording industry or street hustlers for guidance. What more needs to be said? The ironic thing is that, according to statistics, after marriage these same young parents go back to religion in order to provide "moral guidance" for the next generation, and the vicious cycle is continued.

The clergy is constantly badgering "backslidden fathers" with guilt-filled pleas that, for the sake of their children and society, they return to God and religion. The idea that the clergy would have to beg fathers to attend a service if the Almighty God were truly present in their rituals is absurd. These same men will camp out for days to buy a ticket to see the prowess of their fellow humans as they participate in mundane events like the World Series or a Rolling Stones concert. Can you imagine how difficult it would be to score a ticket to church if the Almighty God, Creator of the Universe, were making an appearance? No, it is the total absence of supernatural power in these rituals that alienates fathers.

As if the religious assault on the family wasn't sufficiently destructive, the educational system has recently decided that parents lack the qualifications to fulfill one of their primary duties—the sex education of their children. Therefore, like the clerics, educators are demanding that we abdicate our parental responsibility and turn this personal and morally sensitive function over to amoral science and health teachers. Have they lost their minds? To insist that human sexuality, which floats on a sea of morality, must be taught in an amoral environment reveals the scope of academia's arrogance and confusion. My wife describes these "morally neutral," coed sex-education classes as a brutalization of innocent children. Teachers crassly desensitize pre-teens to the most personal form of human intercourse in a mixed-sex classroom filled with snickering, embarrassed little children being taught how to fit condoms on cucumbers. In the process, they further the degradation of the family by implying that children should not trust the advice of their parents on important matters like sexuality. This gives these vulnerable students the impression that their mothers and fathers are incompetent buffoons who are so afraid of their own sexuality as to be incapable of advising their children. Further, it implies that, providing school safety rules are followed, the formerly sacrosanct marriage bed is merely another fixture on the school playground. Is it possible to conjure up a policy that would be more destructive to both the family and society?

To suggest that this power grab is a carefully thought-out attempt to educate our children into the glorious wonders of human sexuality is a sham. No, this is but one more reckless assault by secular humanists in their ongoing ideological jihad against Judeo-Christianity. It began with an attempt by Planned Parenthood to drive Christian influence from the nation's family life because of the psychological damage incurred by the "suppressed sex drive" of believing parents. It is true that the early Christian doctrines on sex were repressive; however, they were nowhere near as destructive as the current secular humanist backlash. The voluntary Christian strictures on sex may indeed have left us with more than a few uptight, horny and (in some cases) unstable parents. What they did not produce was an epidemic of teenage pregnancies, abortions, venereal diseases and daily scandals from sexually perverted leaders in government, sports, and industry. As one who has lived under both the Christian sexual mores and the secular humanist liberality, I must say that I find the latter to be far more intrusive. It is true that Christianity did require us to voluntarily adhere to certain moral standards, but they never invaded our bedrooms as was slanderously suggested by the humanists, nor did they demand that we relinquish our role as parents and give them full authority to indoctrinate our children into an amoral sexuality. I have had to withstand several attempts to invade the sanctity of my family by left-wing do-gooders in the government. The long-range consequences of this mindless backlash will be far more destructive to society than the repressed sexuality of their equally mindless counterparts in the religious right.

We should all be outraged by this extremely destructive academic assault on the credibility of parents. I remember how offended my children's sex-ed teachers were when my wife and I refused to submit to their intimidating efforts to convince us that our nonconformity would place our children in mortal danger. Let me make it clear that I would have no problem with teaching children the biological functions of the human body in High School, providing they were taught in a single-gender, morally-based environment; but to insist that teachers be allowed to force-feed pre-teens the intimate details of human sexuality in an amoral, coed forum is a perversion of the word education. Even the words chosen to describe this curriculum are an indictment of the current system. There is a vast difference between the connotation implied by the term *sex education*, which requires a moral constraint, and *biology*, which does not. I think there can be no better example of the failure of academia's amoral approach to sexuality than the case of the New Jersey teenager who left the dance floor of her High School prom, gave birth in the women's restroom, dumped the baby into a garbage can and rejoined her classmates at the dance. Please, folks, don't blame this desensitized, coeducated, amoral child; we created her. Are we now going to punish her for being precisely the kind of emotionally detached, amoral person the school system is designed to produce?

There are three institutions charged with the assignment of creating a wholesome society: the family, religion and academia. Clearly, they have failed us in that task, and it is now essential that we find out why and make the necessary corrections. Throughout the history of Western civilization, the Bible was the foundation on which we based our morals. In recent times, the confused and illogical doctrines of Judeo-Christianity, exacerbated by the indoctrination of our schoolchildren into the myth of evolution, have denigrated the Bible, rendering it useless as a source of wisdom for fathers. Convinced that it is impossible for them to solve this conflict of ideologies for their children, modern fathers have abandoned their role in the family, leaving the moral upbringing of the children to their wives and the clergy. The sign of a good father in our modern culture is one who takes his son to a baseball game. That turn of events has been great for the sports business, but disastrous for families. In Eastern cultures, where no such conflict took place, fathers can still turn to Confucius or Buddha for wisdom. To whom can Western fathers turn—the moguls in the sports industry or Hollywood? One of my friends, who has had several nervous breakdowns related to his Catholic upbringing, literally refers his children to the John Lennon song "Imagine" and suggests that, as the song says, they try to imagine that there is no heaven or hell. Another friend, whose religious guilt drove him to alcoholism, escaped by joining the cult of Alcoholics Anonymous. He tells his children that they are nothing more than animals and that out in the world they must eat or be eaten. His wisdom goes on to encourage them to live "One Day at a Time."

Academia's inability to reach a consensus on the salient questions of human existence has left Western fathers confused, thereby reducing some to apathetic breadwinners and others to fanatic advocates for one of the secular or religious ideologies. The fallout from that loss of fatherly leadership in the family is the central cause of the breakdown in Western society. If we are to have any hope of reversing our current state of moral decay, we will have to restore the father to his proper role as the philosophical leader of the family and give him the wisdom to logically answer the important questions concerning life, death and God for his children. We will have to stop the cults of science and religion from teaching these youngsters unfounded doctrines such as prayer, forgiveness of sins, the evil nature of man, amoral sexuality and the evolution of humans from the primates. Instead, we must establish a new code of morals based on the dignity and goodness found in all men. Jesus, who gave us the earth, spoke the words of wisdom that should be the essence of that code when he said,

> Therefore all things whatsoever ye would that men
> should do to you, do ye also to them: for this is the law
> and the prophets. (Mat. 7:12)

The time has come to determine beyond doubt whether the Scriptures are a handbook for modern religion or a closed history of God's dealings with the descendants of Abraham in the ancient world. Though it seems incongruous with their position, many liberal clergymen (whose religions are supposedly based on the Scriptures) have now joined scientists in declaring the Bible to be myth. They claim that miracles such as the parting of the Red Sea by Moses and the resurrection of Jesus from the dead are mythical stories that served as simplistic stepping stones, enabling the immature early believers to come to the reality of our modern-day religions. In contrast, I hold that the reality of biblical history has led us to fabricate our current religious myths. The priests, rabbis and ministers of these powerless religions are forced to declare the biblical miracles to be nothing more than fables in order to avoid being asked to produce a supernatural sign as proof of their divine commission. The truth is that we have the myth and ancient Israel had the reality. It is time to bring the Bible back into the world of academia where it should be openly and respectfully debated like any other historical text. An in-depth knowledge of that history will free us from our modern-day scientific and religious myths and restore the dignity of the father in the family by empowering him to guide his children in their quest for truth through the exercise of their intellects, rather than blind belief.

THE ABDICATION OF WOMEN

A radio talk-show in New York City did a segment on an article in the March 1998 edition of *Glamour* magazine, titled "How Many Men Have You Slept With?" The host invited women callers to comment on the number of sexual partners they had experienced and to state whether they wished the number were higher or lower. Many women proudly admitted that they had had numerous sexual relationships, with the numbers ranging from two to as many as fifty. Most stated that they were comfortable with their number of partners, but a few wished it were lower.

At one point a male caller named Bob said that he had one word to describe the women callers—"sluts." The host immediately began to attack his position as being too judgmental and asked if a woman had had only two partners would he consider her a slut? This poor guy fell right into the trap of trying to decide how many partners would justify declaring a woman to be a slut, and the host quickly made him look like a woman-hating religious fundamentalist. Nevertheless, "slut" is a legitimate word that can be found in the latest dictionary, and its definition (a sexually promiscuous woman) is precisely applicable to the self-described conduct of the women who called the show.

Another female caller asked the host about his own sexual prowess. Sounding somewhat embarrassed, he was forced to admit that his number was much lower than that of most of the women who had called the show. He then proceeded to excuse this shortfall by claiming that females have an unfair advantage because they hold the decision-making power in human sexuality. With that statement he cut to the very heart of the matter. Women do indeed hold the decision-making power over the rate of sexual activity in a given society; therefore, by nature, they have always been the moral arbiters of human sexuality. During the 1960s, academia convinced Western women that they could be "liberated" from the unfair restrictions which that role imposed on their freedoms. This was great news for millions of hedonistic young women, who were only too willing to abdicate a position which had required them to use enormous amounts of self-discipline; however, the growing crises of single mothers, venereal diseases and cancers caused by sexual promiscuity will soon put the lie to the concept that it is possible for science to rewrite the laws of nature. As it becomes increasingly difficult to find a disease-free marriage partner, biology, not theology, will move the next generation of women to force science to face reality.[8]

The most interesting caller was an atheist. He chastised the host for his excoriation of fundamentalist Bob and insisted that regardless of the idiosyncrasies of religion, the citizens of a healthy civilization are required to make moral judgments about the behavior of their fellow citizens. He quoted Thomas Jefferson, who said that without a moral citizenry, a democratic government cannot survive, and he suggested that our philosophical confusion would soon bring Western civilization to a confrontation with that reality. I think that time is already here.

MARKETING FORGIVENESS

As the competition for converts intensifies, the ideological cults of science and religion are becoming ever more skilled at marketing their systems of forgiveness to match both the appetites and fears of the masses. In the case of religion, the marketing strategy has been to blame the devil for the immorality of its followers and ease their fear of death by promising them a glorious afterlife in heaven. Secular humanists were also quick to realize that the masses would not subject themselves to a judgmental hierarchy. Therefore, in order to compete with the religious cults, humanists offered a guilt-free lifestyle by simply rejecting the biblical concept of moral absolutes. They had scientists produce protective devices to insulate their followers from the consequences of sexual promiscuity, and miracle cures (forgiveness) for the inevitable disasters. If a secular believer contracted a venereal disease, produced an unwanted pregnancy or became addicted to drugs, medical science was there to forgive the transgression by treating the destructive behavior as an illness. It is not your fault, you are merely the victim of an inherited genetic defect, a wanton disease, parental abuse or a neanderthalic society. They also provided the false hope that a curative treatment, abortion, clean needle program or some form of psychiatric therapy would allow one to escape the negative consequences of an undisciplined lifestyle. In a shameless attempt to co-opt the immortality offered by the theologians, scientists even held out the false hope that they would eventually be able to eliminate disease and provide humanity with eternal life. Believers who bought this lie, like Walt Disney, began to have their bodies frozen at the point of death so that when science achieved this atheistic paradise, the faithful could be thawed and have yet one more ride on Dumbo.

I realize that some of the statements made in this chapter may seem harsh. Friends who have read the book in manuscript form suggested that I leave it out of the final version for fear that people would think I was too judgmental. After careful consideration, I realized that, to the contrary, no words are strong enough to fully reveal the incredibly destructive nature of our current scientific and religious paradigms. Only a personal confrontation with our individual roles in this crisis will induce the humbling catharsis needed to facilitate a paradigm shift in academia; therefore, it is imperative that both clerics and teachers be confronted with the destructive consequences of their ideological jihad. One of the incidents that moved me to resist the temptation to soften my words occurred as I watched the media coverage of the sex scandal involving President Clinton.

AMORAL IDEOLOGY STAINS THE WHITE HOUSE

When the Clinton scandals hit the White House, it became apparent to me that academia's ideological chickens were coming home to roost. Like everyone else, I listened as incredulous commentators told us that despite their graphic exposé on the sordid details of this spectacle, the President's poll ratings continued to soar. Even commentators who are known to lean to the left appeared to be incredulous at this phenomenon. Why were they so shocked? Are they unaware of the fact that, ever since the Scopes trial in 1925, academia has been indoctrinating American schoolchildren with an atheistic system of non-judgmentalism? Were they absent during the elections of 1992 and 1996 when academics assured voters that character didn't count in a presidential election? The real question is, How did our brightest intellects in academia reach such an illogical conclusion? These are the same men who insist that all life emerged from one primeval cell, that the Great Pyramid is a tomb and that there is no evidence of a supernatural intervention in the ancient world.

The possibility that the President of the United States hit on an intern in the White House and then asked her to perjure herself before the bar of justice should be appalling to any normal human being, and, when the scandal first broke, Clinton's poll numbers began to drop precipitously. Alarmed by this logical reaction, the amoral ideologues in academia immediately came on television to warn us that if we allowed any remnant of our primitive morality to rear its ugly head, the French would laugh at us. It is a common tactic for cult leaders to use peer pressure to force people to deny their individual instincts; after all, no one wants to be so unsophisticated as to admit that the emperor is naked. Like teenage children who rearrange their entire lives to avoid being laughed at by their peers, American academics seem to be traumatized by the prospect of inducing French laughter. Are the French so noble that even their sense of humor has become infallible on matters of morality, or does academia's homage have more to do with the fact that the secular humanist philosophy was founded in the city of Paris?

No, folks, this is no joke. We have serious problems, and the President is every bit as much the victim in this case as are his constituents. The advent of an amoral presidency was the inevitable result of allowing the cult of academia to use our schoolchildren as lab mice in a massive sociological experiment. For over half a century, secular humanist ideologues have brainwashed us with their atheistic, non-judgmental philosophy, producing an entire generation of morally confused citizens, and, according to the John Leo column, the prospects for the next generation are even more frightening. We should no more blame Bill Clinton for his amoral presidency than we should blame the public for ignoring his desecration of the office or the "prom-mom" for throwing her baby in the garbage. These are the tragic consequences of the mindless philosophical power struggle in academia.

Let All the Poisons Lurking in the Mud Hatch Out!

Near the end of the ancient Roman Empire, infighting among the differing factions in the morally decadent aristocracy became vicious. All sense of public good was lost as the opposing forces used bribery, deception and murder to achieve power. Repulsed by this depravity, Emperor Claudius wanted to do away with the monarchy and install a republic. Claudius had two sons: Britannicus, a mild-mannered young man, and a stepson named Nero, who was a complete madman. With great intrigue, Nero's mother, Agrippina, was planning to kill both the Emperor and Britannicus in a scheme to bring her son to the throne ahead of the legitimate heir. Claudius reasoned that if he allowed Agrippina to achieve her goals, the emperorship of the demented Nero would expose the perverse nature of the entire system and force the citizens of Rome to make a change.

In a similar manner, the ideological machinations of the cult of academia will soon confront us with the reality that it is every bit as dangerous to have an amoral ideologue as President as it was to have a Born Again Christian. That clash of ideologies has left us incapable of making the most basic judgments, even as to what is or is not proper behavior in the Oval Office of the White House. While the French may see that as a sign of our maturity, I see it as evidence that academia is still lost in the deep ignorance of the dark ages, unable to tell us with any certainty who we are or how we should behave. As the success of Agrippina's conspiracy exposed the corrupt nature of the Roman aristocracy, the success of the secular humanist schemes to elect an amoral President are exposing the ideological poisons that have been lurking in academia's philosophical mud for generations. If we are to eliminate the disastrous ethical, economic and military consequences of these radical swings in ideology, we must challenge the hierarchies in science and theology to solve the mystery of our existence and insist that their next paradigm be based in fact and truth.

VII. NO PLACE TO HIDE

THE RAPID DEVELOPMENT OF TECHNOLOGY OVER THE LAST century is creating a virtual global village. Our expanding means of travel and communication are making it increasingly difficult to ignore the contradictory philosophical beliefs of our fellow man. Television, radio, newspapers, movies and the internet are bringing foreign cultures into our lives and, in the process, revealing us to ourselves. It seems that what we have failed to notice in our personal beliefs is quite easy to see in the unusual rituals of others. Being exposed to these divergent religions is slowly forcing us to reexamine the complex and bizarre concepts we humans have of God and the origins of man. Eventually, these inter-cultural relations will force us to admit that we can't all be right.

On Friday, April 18, 1997, ABC's "Nightline" aired a television program called "The Hajj," which told the story of the Muslim pilgrimage to Mecca. ABC reporter Michael Wolfe, a convert to Islam whose mother is a Christian and father is a Jew, took a camera crew with him on his journey to the site of the most holy shrine in Islam. For the first time, non-Muslims were allowed to view this religious phenomenon, in which millions of believers from around the world make the pilgrimage in a virtual sea of humanity. The Hajj begins with a cleansing ceremony in which participants strip themselves of all earthly possessions, wrap themselves in two plain white sheets and don a pair of simple sandals for the journey. When the pilgrims reach the site of the Ka'ba (the house of God first built by Abraham) several thousand at a time move through the temple in a circular motion and either kiss or raise their hands to the black stone which is thought to be the only remnant of the original building.

Naturally, the logistics of safely moving such vast numbers of people through the ceremony are mind-boggling, and often the crush of fervent believers leads to disastrous stampedes and tramplings with numerous casualties. This year was no exception, and during Wolfe's visit, a fire in the Indian quarters left approximately 350 people trampled or burned to death and 1,300 injured. Like a nervous public relations agent trying to reassure people of the reliability of his client, one Muslim frenetically tried to ease the implications of this tragedy by saying that it would be a great honor to die for Allah on such a holy occasion. His statement brought to mind Billy Graham's efforts to justify God's failure to save the children in the Oklahoma City bombing. All those interviewed by Wolfe said that the Hajj was a spiritually moving experience that would live with them for the rest of their lives. Ted Koppel closed the show by stating that perhaps we have more in common than we realize.

Mr. Koppel is absolutely right; in many ways, we are more alike than we had previously thought. Obviously, we are all seeking to understand and obey the wishes of God, yet the clergy differ greatly over how that should be achieved. Consequently, many people are confused as to what religion to follow. This television program was a classic example of how technology is familiarizing us with our diverse religious beliefs and, in the process, making us less fearful of each other. Years ago, we rarely interacted with people from different faiths, but this year alone my teenage daughter, whose best friend is a Muslim, received numerous invitations to bar mitzvahs and confirmations. Also, some close friends invited us to a Muslim wedding service in their home, and some new neighbors asked us to participate in a Hindu ritual to bless their family. Only a generation ago, my Catholic parents would have refused to participate in any of these rituals on the grounds that the Church considered them satanic. That is a dramatic change for the better in one generation. In fact, Mr. Wolfe's parents are a prime example of this unavoidable intermingling; although I am sure it is not easy to accommodate the dictates of Christianity, Islam and Judaism in one family. Clearly, it is growing increasingly difficult to ignore our differences. At some point, inter-religious marriage, international trade and the technologically shrinking planet will force us to try to reach an accommodation. To do that we will have to ask ourselves some difficult questions.

Ask The Right Question

A basic premise of science is that to find a truth we must first ask the right question. Often the key questions are readily apparent but politically incorrect, so they remain unasked. Unfortunately, the philosophical crisis in science and religion has brought us to a place where the following inquiries are simply unavoidable.

The Scientists

To the scientific believers, I would ask: Why have you gone to such great lengths to solve the mysterious disappearance of the dinosaur while completely ignoring the even more compelling disappearance of the ancient gods? Why has science failed to produce an explanation for the demise of the ancient world, the great loss of knowledge and the subsequent rise of man to rule the planet? Why have you discounted the voluminous ancient texts from around the world that uniformly declare man to be an alien while spending multiplied millions of dollars trying to prove the contention of a single man that you are a relative of the primates?

Why do you doggedly persist in an obviously futile quest to conjure up a scenario that will authenticate evolution when it has now been proven to be a myth? Why have you dismissed out of hand the comforting biblical treatise that man is a unique being created by a loving God who gave him the earth and plans to give him eternal life? Have you fully investigated this thesis and rejected it on the merits, or is ideological dogma preventing you from being objective about biblical history? In light of the total collapse of the Darwinian paradigm, wouldn't it be prudent to reexamine the Bible's account of our creation in Eden and subsequent exile to earth?

ACADEMIA

I ask the academicians: Why have you banned serious study of ancient Jewish history (the Bible and Josephus) from your centers of learning while making the study of ancient Rome and Greece required course work? Greek and Roman history are, for the most part, irrelevant; however, a knowledge of Jewish history is absolutely essential to a proper understanding of modern civilization. It is at the heart of the split between science and religion, the bombing of the World Trade Center, the Left vs. Right political struggles, Bosnia, Belfast and the Middle East. Shouldn't such an important history be part of the curriculum of every school in the Western world? You have driven Jewish history from the educational system, describing it as "a dangerous myth." Since when is the study of history dangerous? Are not the atheistic philosophers who gave us the myth of evolution, the Darwinian Holocaust and our current amoral education system, dangerous? Atheistic zealots in communism and fascism used Darwinian philosophy to get men to follow them to a "bold, new scientific future" that resulted in the slaughter of nearly one hundred million people in Europe alone. Yet you continue to indoctrinate our children with this obnoxious myth while describing biblical history as dangerous. Is it not dangerous to allow people who describe themselves as "amoral atheists" to teach our children, and is it realistic to expect such a cynical education to produce a stable young adult? Can one be an anthropologist and have no knowledge of the fall of the gods and subsequent rise of man to rule the earth? Is it possible to be a historian without knowing the history of the ancient Israelites? Is it realistic to claim to be an archeologist and be unable to explain the collapse of the ancient world? Would the ancient world system of slavery have been carried over into the modern world if you had properly researched the fall of the gods? Would there have been a civil war in America? By adopting the theory of evolution, you endorsed Darwin's position that the blacks were savages who had not yet fully evolved from primate to human. Would it be unfair to suggest that that endorsement gave southern slave-owners an air of intellectual paternalism that prolonged the system of slavery? Would the African Americans be as angry at their fellow man if you had taught them the true history of their enslavement, instead of blaming it on the "white man?" Is it not true that the first attempts to overthrow the ancient system of slavery were initiated by revolutionary governments created by white men in Western civilizations? If, as educators claim, modern racism is the product of ignorance, whom should we blame for the ignorance of these racists? Would it have required violent revolution to overthrow the ancient monarchies or taken so long to develop a human government if the intelligentsia had followed the historicism of Hegel rather than the radical atheism of the Empiricists? Could the devastating ideological wars of Europe, including the Holocaust, have been avoided if you had understood the wonderful truth inherent in the fall of the gods, collapse of the ancient world and subsequent rise of man to rule the earth? Finally, it is the primary duty of academia to solve the mysteries of human existence. If you were asked to give yourselves a grade for your efforts to reveal the origins, purpose and future of mankind, what would it be?

THE JEWS

To the "monotheistic" Jews, I ask: Did the one God of Abraham produce more than one truth? If there is only one God, there can only be one truth about Him. Why then are there several denominations of Judaism based exclusively on the Old Testament?[1] Can it be that God speaks with a forked tongue, or are the rabbis listening with forked ears? In light of the fact that rabbis are unable to agree on the facts contained in the Old Testament, how credible is their ability to be unanimous in their rejection of the New Testament, which they refuse to study?

The New Testament was written exclusively by and about Jews; doesn't that make it Jewish history? If the answer to that question is yes, and it obviously is, how can rabbis reject their own history? Do they ban the New Testament from their synagogues because they think that Jesus was a false Messiah? Why is Jesus censored, while all the other mistaken Messiahs of Jewish history are required course study? Should we now expect the orthodox Jews who recently thought that Rabbi Schneerson was Messiah to ban his writings and pretend that he is not part of their history? Jesus is the most important Jew ever to walk this earth. His life has had a dramatic impact on both the ancient and modern worlds; yet, for the most part, rabbis refuse to acknowledge his existence. To claim to be a Jewish scholar and fail to recognize Jesus is the equivalent of claiming to be a student of American history while refusing to acknowledge the life and work of Abraham Lincoln.

The entire history of God's dealing with the ancient Israelites belongs to the Jews. It is just as ludicrous for rabbis to reject the New Testament portion of that history as it is for Christians to claim it. Finally, the Old Testament Scriptures prophesy that in the end times a civil war among the Jews[2] would bring an end to the daily sacrifice,[3] destroy the temple and Jerusalem, and that the Spirit of God would leave the earth.[4] In A.D. 30, Jesus told the Jews of his day that they would live to see the fulfillment of those prophecies.[5] In A.D. 70, the temple and Jerusalem were literally annihilated and the Ark of the Covenant, the very manifestation of God's presence in Israel, vanished without a trace. What message was the God of Abraham sending to the ancient Israelites?

THE CHRISTIANS

Finally, my questions to the Christians. For nearly two thousand years Christian leaders have used the New Testament end of the world prophecies to terrorize mankind with the prospect that God was about to destroy the planet. In light of the fact that all of these prophecies were written prior to the demise of the ancient world, is it unreasonable to assume that they refer to that cataclysm and, therefore, hold no threat for modern man? In the forty years between the coming of Messiah and the fall of Jerusalem, twenty-seven books of the New Testament, filled with miracles, were written. Christianity claims that the Bible is an ongoing revelation; yet in the nineteen hundred years since, not one word or miracle has been added to this book. Isn't it time for those of us who were indoctrinated into Christianity to admit that the lack of supernatural intervention by God in the modern world is undeniable proof that the Bible is a closed history of God's dealings with the ancient Israelites?

Like the Jews, you would be well-advised to keep in mind the concept that one God implies one truth; yet there are literally thousands of denominations of Christianity, each claiming to be the exclusive conduit to God and the afterlife. Instead of dividing yourselves into hostile camps of saved and unsaved believers, wouldn't we all be better advised to seek common ground? Since we can't all be right, what would be so terrible about coming to grips with the possibility that we are all wrong? What would be the harm in seeking new and less threatening answers to our questions about God, life, death and the afterlife? In light of the fact that many of your past religious beliefs have proven to be wrought with error, is it wise to continue to kill one another over your current beliefs while adamantly rejecting any new knowledge on these all-important subjects? The solar system is a very precise and logical creation, therefore we must assume that its Creator is a logical being who gave man intellect to enable us to understand Him and His creation. Do we not honor God and show great faith in His goodness when we use that intelligence to seek the truth? Keep in mind that there is a vast difference between questioning God and questioning what we believe about Him.

"GIVE ME A CHILD UNTIL HE IS SEVEN, AND I HAVE HIM FOR LIFE" —KARL MARX

Though Mr. Marx's theories on economics have been largely discredited, this statement is as valid today as the day it was first offered. Most people have very strong beliefs that are not really their own. We have all heard expressions like: "I was born a Catholic and I'll die a Catholic"; "I was born a Jew and I'll die a Jew"; "My father was a union man like his father before him"; "My parents are Democrats, so I'm a Democrat." Most of our beliefs are inherited, and they were imprinted on our psyche early in life. Consequently, no matter how illogical they may be, we will defend them to the death. We connect these beliefs with either God or our parents, and when their veracity is challenged, we react as though the very essence of our existence, or God Himself, has been attacked.

The leaders of our secular and religious cults spend most of their time trying to find ways to reinforce those early indoctrinations. They do all that they can to "protect the flock" from outside influences by creating fear of others. Christians consider the doctrines of opposing denominations to be satanic and warn their members not to associate with these apostate brethren for fear of "losing one's faith." Orthodox rabbis throughout the world are attempting to decertify the ordination of non-orthodox rabbis and declare the religious ceremonies of all the other denominations of Judaism to be null and void. Meanwhile, academia bans biblical history from the schools on constitutional grounds. All of these efforts to isolate their followers from external input can be very effective in keeping the sheep from straying too far from the ideological fold.

As a child in Catholic school, I was taught that to so much as doubt my faith would place me in danger of eternal damnation. I was told to drive such thoughts from my mind and avoid people with conflicting beliefs. That early indoctrination made me afraid to trust my own judgment. I was unable to listen to new ideas, question early beliefs, see a movie or associate with anyone not approved by my church. Most of my adult life was spent bowing to "professionals." I wouldn't dare to question the clergy, schoolteachers or even the instructions of my doctor. When I developed my theory and began to challenge these professionals, all of that changed. I had looked death and the false god of religion in the eye, and they had both blinked. I began to realize that my thoughts were just as valid as those of the experts, and suddenly I was a truly free man.

It is impossible to explain to someone who is still bound by religious beliefs the release one experiences when those psychological shackles are thrown off. Once that mind-set is broken, a great freedom rushes into the void. Sunday became a day dedicated to my family, rather than to church services. Gone was the frustration of sitting through endless rituals and quietly submitting to the insatiable demands of religious doctrines. No longer would I be required to eat certain foods or wear certain clothes. Never again would I allow guilt to motivate my actions. No longer did I feel driven to testify to the "lost" members of my family and friends. Guilt-ridden Christian holidays became happy vacations, and when the false god of religion was driven from my marriage bed, I was able to enjoy sex without feeling dirty or sinful. Fear of death became anticipation of what the second life would be like, and fear of God became awe for His genius. As a Born Again Christian, I was required to give ten percent of my income to the church. Now my income was fully mine. For the first time in my life I had the courage to think for myself. No longer able to believe that prayer would entice God to help me solve my problems, I was forced to take full responsibility for my life. My thinking became concise, and I saw old problems with a new objectivity. Unlike the members of my family who have left the Church but are loaded with guilt, or friends who remained but were forced to become philosophical hybrids to accommodate the evolutionary indoctrination of their children, I walked away from these conflicting paradigms without the slightest pang of conscience. I felt fully capable of guiding my children to a knowledge of the truth and confident that it would make them immune to the seductive lure of the academic and religious cults that have mentally enslaved their peers.

> And you shall know the truth, and the truth shall
> make you free. (John 8:31)

As a Christian, I had heard these words of Jesus quoted many times. Now I not only understand their meaning, but have experienced their reality.

THERE IS NO SANTA CLAUS!

Sometimes, in an unguarded moment, a thought crosses our minds that completely contradicts our current beliefs. At first, we may toy with the new concept, but when we start to think of the consequences of changing our philosophy on the subject—admitting we were wrong, enduring family pressure etc.—we often decide that it is just not worth the effort. A similar situation occurred when I told my sister Susan about my theories. At first she was very curious and asked to read the manuscript, but as she left my home with the work, I could see she was having second thoughts. Susan is a Born Again Christian, and when the read was finished, she wrote me a terse note saying that, though the theory was compelling, she felt bullied by the tenor of the writing. She closed by saying, "people will hate you for it."

Susan's prediction that people would hate me for revealing the truth reminded me of a similar experience I had as a child. When I was about nine or ten years old, my older brother Danny took me aside and told me that there was no Santa Claus. Danny justified his revelation by saying that the family's finances were making it impractical for me to continue to believe in Santa and that I was now old enough to join my older siblings who already knew the truth. Though I felt privileged to be joining the adults, I hated him for destroying a fantasy that I had grown to love. It wasn't that Danny had told me something I didn't already know; he just made it impossible for me to continue the pretense. That incident spoiled our relationship for many years, and poor Danny had to take the heat for exposing a hoax he didn't perpetrate. I guess there is no nice way to tell someone that they're wrong or that they have been deceived, even when they already suspect as much. It was comforting to believe in Santa as a child, but eventually all myths become impractical and we are forced to face reality.

Our Naked Emperors

In the Hans Christian Andersen fable "The Emperor's New Suit Of Clothes," a monarch is led to believe that he has an invisible suit. In reality, he is naked, yet no one in the empire is willing to admit that they cannot see the clothes. Instead, each tries to outdo their fellow subjects with flatteries for the nonexistent garments. During the emperor's first public appearance in the new suit, a little child boldly states that he is naked. Immediately, both the crowd and the emperor realize that they have been deceived; however, just as quickly, they decide to continue the pretense in order to avoid an embarrassment, and the royal procession moves on with the adults continuing to praise the emperor's new clothes. In real life, as in the fairy tale, when our erroneous beliefs clash with reality, peer pressure can move large numbers of people to jointly deny an inconvenient revelation of the truth.

A classic case of group denial was broadcast to a worldwide audience during the funeral for Diana, Princess of Wales. The traumatic life and death of the Princess is stark testimony to the destructive effects of trying to maintain a pretense of ancient royalty here in the modern world. Throughout the broadcast, commentators competed with each other for the smallest details on the royal family, yet never mentioned was the meaning of the word *royal*. The term *royal bloodline* is derived from the ancient system of primogeniture and implies that the members of the British monarchy are descendants of the gods. Like the Roman emperors after the collapse of the ancient world, the British monarchy is expected to affect an air of being more than human. In order to maintain that fantasy, they must keep the public at a distance, lest they become too familiar and spoil the illusion. Unlike Prince Charles, Diana had spent much of her early life mingling with normal society. This did not prepare her for the almost monastic life of the royals. Her difficulty in adjusting was a clear case of the aloof stoicism feigned by the monarchy being incapable of suppressing the vivacious human spirit of this wonderful young woman. Unfortunately, her boundless humanity and simple honesty clashed with the need to affect a sense of majesty in the royal family and eventually led to her divorce, loss of title and tragic death. Like the adults in the Hans Christian Anderson fairy tale, the billions of us who watched her funeral on television already knew that Diana had been spiritually dying for years under the demands of this forced pretense, yet we were never able to bring ourselves to forgo the cherished myth that perhaps she was a goddess. For the sake of her children and the mental stability of the rest of the "royals," the people of England should carefully weigh the consequences of keeping these people imprisoned in this anachronistic myth.[6]

As an aggressive media is making it increasingly difficult for the members of the monarchy to maintain the illusion of royalty, technology is also making it impossible to maintain the myth that man evolved from the primates or that we can communicate with the Almighty God. Here again, most of us already know that these beliefs are unfounded; however, like the childhood myth of Santa and our fascination with British pageantry, we just don't want to know for certain. Nevertheless, at some point a combination of human progress and philosophical terrorism will force us to think the unthinkable and admit what, in our hearts, we already know: that our scientific, religious and royal emperors are naked!

I take no pleasure in upsetting the beliefs of others, but there is simply no place to hide. It is not my work that is forcing us to confront our erroneous beliefs, it is the technological revolution. That inexorable march of human progress is causing us to intermingle with each other, thereby exposing the absurdity of our respective beliefs. My discovery is merely a part of the fallout from that reality. It is the product of a difficult struggle for both me and my family, spanning a period of more than fifteen years. I am convinced that it is the truth and that when it is fully understood, we will all delight ourselves in the abundance of peace prophesied by King David. To that end, I have started an organization called The Realists, which is dedicated to a pragmatic reexamination of ancient history, both biblical and pagan, for the purpose of solving the great mystery of man's existence on planet Earth. Our motto is One Man, One God, One Truth.

www.realists.org

Endnotes

Introduction

1.   This term will be used to describe early Christianity, which later took the name "Universal" or "Catholic" Church.

Chapter 1

1.   This is used as a blanket term to describe all religions that are based on the history of God's dealing with the ancient Israelites as recorded in the Old and New Testaments of the Bible. I include Judaism, Christianity and Islam in that classification.

2.   The period of history that begins with the fall of Jerusalem in A.D. 70.

3.   Throughout the book I will refer to the atheists and secular humanists as atheistic, empirical or secular scientists, academics or intellectuals; on the other side, for the most part, I refer to the clerics as theologians, religious leaders and clergymen.

4.   Eldredge, N. and Gould, S. J. (1973) "Punctuated Equilibria: An AlternativeTo Phyletic Gradualism," in *Models in Paleobiology,* ed T.J.M. Schopf, Freeman, Cooper and Co, San Francisco, p. 82-115.

5.   A series of circular motions superimposed on each other.

6. The concept that all combustible matter contained an element called phlogiston, which caused combustion.

7. II Tim. 3:7.

Chapter ii

1. The remains of animal and human sacrifices and large upright pine posts have been found in these pits.

2. *In Search Of Ancient Gods*, p142-43 & 236-37.

3. *In Search Of Ancient Gods*, p114-15 & 359.

4. This hollowing, barely visible to the naked eye, went undetected for centuries.

5. Defined as the fundamental principles that order the universe and shape human nature.

6. *In Search Of Ancient Gods*.

7. Catalogued as the Akbar-Ezzeman MS. by Abu'l Hassan Ma'sudi.

8. Distinguished professor of the History of Science, whose research into the advanced science of geography, geodesy and astronomy evinced by the ancient Egyptians and Sumerians spans more than thirty years.

9. *Gods* with a small *g* refers to the pagan gods.

10. A demigod was the offspring of a union between one of the gods and man; a mortal with some of the attributes of a god.

11. After speaking with God, the face of Moses shone with such brightness that he had to cover it before speaking to the Israelites. (Ex. 34:29-35)

12. The *Ramayana* and *Mahabharata* of India, and *Beowulf* of the Nordic people, tell similar tales of interactions between the gods and men.

13. The great Chinese writer Confucius (551-458 B.C.) said, "I am not

an originator, but a transmitter."

14. This was not merely a snake. The bible describes this creature as a very subtle entity, not something that would be easily understood by modern man.

15. See Gen. 3:1-24.

16. Gen. 3:14-15, Rom. 16:20, Mat. 5:5.

17. The Hebrew word translates as "The anointed of God."

18. This was the wooden staff that God made to bloom and bring forth almonds as a sign that He had chosen Aaron to serve Him in the tabernacle of the Lord. (See Num. 17:10 & Heb. 9:4)

19. Manna is the food that fell from heaven to feed the Israelites during the forty years they wandered in the desert after leaving slavery in Egypt. (See Ex. 16:33-34 & Heb. 9:4)

20. See II Chr. 6:41.

21. See Josh. 4:5-11&18, 6:6-20; I Sam. 5:1-12.

22. See II Chr. 36:6-18.

23. Jer. 32:1-44 & 33:1-26.

24. A Christian term for the satanic counterpart to the Messiah.

25. See the Old Testament Book of Nehemiah and Jos. Bk. 15, 11:1.

26. See Mat. 21:13.

27. Josephus gives a detailed description of this work describing the stones used in the construction. "Now the Temple was built of stones that were white and strong, and each of their length was twenty-five cubits, their height was eight, and their breadth about twelve." (Bk. 15, 11:2-3) A cubit is 21 inches long, making these stones 43.75' x 14' x 21'.

28. See Mat. 2:1-12, Num. 24:17 & Is 60:3.

29. These doors were made of solid brass and could barely be moved

by twenty men. They sat on an iron plate and were locked with heavy iron bolts that sank deep into the temple floor. (Jos. Bk. 6, 5:3)

30. See also: Hosea 1:11, 6:1-2; Jer. 23:5-8; 1 Thes. 4:13-18; Mat. 24:30; 2 Cor. 15.23, 15:52.

31. See also: Mic. 7:1-13, Mat. 10:21, Luke 12:52-53.

32. See also: Mark 13:12 & Mat. 10:21.

33. See Jos. Bk. 4, 7:3.

34. It was a great sacrilege for a Jew to remain unburied.

35. Son of Emperor Vespasian.

36. This was a fulfillment of prophecy. See Dan. 9-12.

37. Josephus was a high priest, a general of the army and a prophet of God.

38. See Mic. 7:1-7.

39. Sedition: language or conduct directed against public order and the safety of the state. Seditious: those who engage in sedition.

40. See Ezek. 10:2; Is. 66:15-24, 27:1, 2 Pet. 3:10-12, 2 Thes. 1:6-10.

41. See Jer. 26:18, Mic. 3:12 & Mat. 24:2.

42. Jos. Bk. 6, 3 (p.577).

43. Jos. Bk. 6, 4:1-7 (p.579-580).

44. See Jos. Bk. 6, 4:5, p.580 & Ezek. 10:2.

45. The term *high places* in the Bible refers to temples.

46. See Jos. Bk. 6, 5:3 (p.582); Tacitus, History 5:11-13 & Ezek. 10:5.

47. Rev. 19:20 & 20:10.

48. See Josh. 4:5-11&18, 6:6-20; I Chr. 13:9-10; I Sam. 5&6.

49. Jos. Bk. 5, 11:2 (p. 565).

50. *Raiders Of The Lost Ark*

51. See Ezek. 10:1-22, Jer. 3:16 & Rev. 11:19.

52. See Gen. 3:1-24.

53. See Rom. 16:20 & Mat. 5:5.

54. The Greek word translated *meek* in this verse refers to the humans, who were the lowly beings of the ancient world.

55. The Castle is 78 feet high with a staircase of 91 steps on each side. The total number of stairs including the upper platform allows the pyramid to harmonize with the 365 days of the solar year.

56. It is believed that the heads of the players who lost on the ballcourt were used in rituals performed on this platform.

57. Anyone who has read of the mortal combat between animals and humans in the Roman amphitheaters will have no trouble imagining the rituals performed on this platform.

CHAPTER III

1. Goliath was 9 feet, 9 inches tall. He wore a brass helmet and a bronze coat that weighed 166 pounds. His spear was huge, the head alone weighing 20 pounds, and two men carried his shield. (1 Sam. 17:1-45) Imhotep in Egypt was equal in stature.

2. The spectacle of these once all-powerful rulers, who could have taken the lives of their subjects for back taxes, being told how much of a stipend they will be given and having that stipend taxed by a bunch of humans in Parliament is prima-facie evidence for my position.

3. With the loss of the ancient economic system, some of the Indians in the Western hemisphere who had served the Inca and Mayan empires were forced to move to the coast where they began to

trade in the gold and silver relics of these lost civilizations. It was that wealth that would later attract Spanish explorers like Columbus, Pizarro and Cortés to the continents we now call the Americas.

4.   Interim king.

5.   *A History Of Rome*, Ch. 5:5-6.

6.   See Josephus, *Wars of the Jews* Bk. 3, 8:9 p.516; Bk. 4, 10:1-7 p.543-5; Bk. 6, .5:4 p.583.

7.   See Ps. 37:11 &37:37, A History Of Rome, Ch. 36:1&4

8.   Vespasian's two sons, Titus and Domitian, succeeded him as emperors of Rome.

9.   Spanning the years A.D. 69 - 180.

10. *A History of Rome* p.483.

11. The supreme council and highest court of ancient Israel.

12. Mat. 24:36-51, John 12:48.

13. Note the difference between the way God protected the Jews in the Babylonian captivity and the way He left them to the hands of the Romans after the fall of Jerusalem in A.D.70.

14. M. Ulpius Traianus, Roman Emperor (98-117).

15. A Jewish friend told me the story of an incident that occurred during his military service in WWII. As a young man, he was shipped to Europe. On the way over, Irving noticed a couple of sailors from Tennessee looking at him with more than normal curiosity. After a few days, they finally approached and asked if he were Jewish. When he answered in the affirmative, they asked to see his horns. He tried to assure them that Jews did not come with horns, but they would not believe him until he allowed them to personally inspect his disappointingly normal skull.

16. This refers to the Israelites of the ten tribes of the Northern Kingdom, not to non-Israelites, as the Church teaches.

17. In fact, they coined the term *Dark Ages* and deliberately limited this time span to the period between A.D. 476 and the end of the 10th century. This ignored the deep ignorance of the period between A.D. 70 and 476, but enabled them to lay the blame for this intellectual dearth on the Church.

18. This erroneous doctrine was based on a misinterpretation of the verse in Genesis.

19. This title came later.

20. The dynamics of the planet (atmospheric, geological etc.) are now as they always have been.

21. Natural Selection: The theory that variations advantageous to an organism in a certain environment tend to become perpetuated in later generations as part of the concept of the survival of the fittest.

22. 130 years later, scientists are still seeking that phantom.

23. In Rudyard Kipling's story "The Man Who Would Be King," the hero was able to dominate a nation of superstitious savages until one day he was wounded and the tribesmen saw blood flowing from his body. Immediately they cried, "Not gods, not gods; men! men!" This truly was the situation in Europe.

24. A bundle of birch rods fastened to an ax.

25. The fact that Marx, a German Jew whose father had him baptized, rejected the German school of philosophy for the empiricism of Paris proved to be a contributing factor in Hitler's rabid anti-communism and anti-Semitism.

26. The theory that scientists could improve society by methodically managing the human race.

27. Jews wore the star of David; homosexuals, pink triangles.

28. *Michael, Michael, Why Do You Hate Me?* p.33, by Rev. Michael Esses, former Orthodox rabbi.

29. Gen. 38:9.

30. Colin Ferguson's attack on white commuters on New York's Long Island Rail Road.

Chapter IV

1. Mat. 12:25.

2. The rapid growth of the Islamic population has added a whole new dimension to this problem.

3. The Unification Church claims that Sun Yung Moon is the fulfillment of the prophesied second coming, while the Ahmaddiya sect of Islam, founded in India in the late 19th century, believe that their leader, Hazraat Mirza Ghulam Abmad, is the Messiah.

4. Ancient Jewish religious sect.

5. See also John 12:19.

6. See Gen. 17:1-19, 15:1-16 & Deut. 7:6-7.

7. See Heb. 8:8-12.

8. See Heb. 8:1-13.

9. See Rom. 11 & Acts 10.

Chapter V

1. My disparaging remarks about the doctrines of Judeo-Christianity have led the clergy to accuse me of "making fun of other people's religion," as though that were some kind of heinous offense. The account of Elijah's gibes at the impotence of the prophets of Ba'-al should put that complaint to rest.

2. See also I Ki. 13:1-6 & Mark 16:17-18.

3. Recently, a tipster enabled the FBI to prevent a terrorist bomb

attack on a key subway terminal in Brooklyn.

CHAPTER VI

1. It is interesting to note that theologians also found it convenient to look the other way when their distraught followers adopted this bifurcated belief.

2. "On Society," *U.S. News & World Report* 7/21/97

3. See Rom. 4:15 & 10:4

4. *Crime and Immorality in the Catholic Church,* p. 77

5. *Crime and Immorality in the Catholic Church,* p. 32.

6. See chapters 12 & 13, *Crime and Immorality in the Catholic Church.*

7. by Agostino Bono, *The Tablet* 11/17/90.

8. It was recently discovered that over sixty percent of the young women on the campus of a university in northern New Jersey carry the virus that causes venereal warts.

CHAPTER VII

1. See I Ki. 18

2. Mic. 7:1-6, 12-13; Dan. 8:11-14, 12:11-13; Ps. 37:9-11, 35-36

3. Dan. 8:11-14, 12:11-13

4. Ezek. 10:1-22

5. Mat. 24:34

6. Keep in mind that Charles (who loves an ineligible woman) only married Diana in order to produce an heir to the throne

Bibliography

Berlin, Isaiah. *Karl Marx,* 4th ed. Oxford: Oxford University Press, 1978.

Cary, M., and H.H. Schullard, *A History Of Rome Down To The Reign Of Constantine,* 3rd ed. New York: St. Martin's Press, 1975.

Denton, Michael. *Evolution: A Theory In Crisis.* Bethesda, Maryland: Adler & Adler, 1986.

Esses, Michael. *Michael, Michael, Why Do You Hate Me?* Plainfield, New Jersey: Logos Intl., 1973.

Holy Bible, Authorized King James Version, The Open Bible edition. Nashville: Thomas Nelson, Publishers, 1975.

Josephus, Flavius. *Josephus Complete Works.* Translated by William Whiston. Grand Rapids, Michigan: Kregel Publications, 1981.

Kushner, Harold S. *When Bad Things Happen To Good People.* New York: Avon Books, 1983.

McLoughlin, Emmett. *Crime And Immorality In The Catholic Church.* New York: Lyle Stuart, 1962.

Pisar, Samuel. *Of Blood And Hope.* Boston: Little, Brown & Co. 1980.

Reed, David A. *Behind The Watchtower Curtain.* Southbridge, Massachusetts: Crowne, 1989.

Robertson, Archibald. *The Origins of Christianity.* New York: International Publishers, 1954.

Rue, Loyal D. *By The Grace Of Guile.* New York: Oxford University Press, 1994.

Strong, James. *Strong's Exhaustive Concordance Of The Bible.* Nashville: Crusade Bible Publishers, Inc.,

Tacitus, Cornelius. *The Annals Of Imperial Rome.* Translated by John Jackson. Cambridge, Massachusetts: Harvard University Press,

Tacitus, Cornelius. *The Histories.* Translated by Clifford H. Moore. Cambridge, Massachusetts: Harvard University Press,

Tompkins, Peter. *Secrets Of The Great Pyramid.* New York: Galahad Books, 1971.

von Daniken, Erich. *Chariots Of The Gods?* London: Souvenir, 1969.

von Daniken, Erich. *In Search Of Ancient Gods.* New York: Putnam, 1973.

World Christian Encyclopedia, Edited by David B. Barrett. Oxford: Oxford University Press, 1998.

Index

A

A.C.L.U. 168, 174, 177–178
A.D. 70 26, 60-69, 88, 91, 105, 110–118, 164, 200, 239, 260
Abortion 174–176, 184, 229–235, 240, 244, 250
Abraham 117, 247
Absolutophobia 225–230
Academics 5, 30, 47, 95, 172, 251
Alien 1–7, 20, 22, 53–56, 103, 131, 185, 257
Allies 134, 139, 141–147, 156, 161, 163, 177–178, 213
Alogical 6–7
America 107, 121, 134–135, 144–145, 161, 165, 167–168, 171–173, 177–184, 191–192, 226–238, 259
Amoral 6, 7, 125, 149, 157, 175–176, 184, 217–232, 243, 245–246, 251–253, 258–259
Ancient gods 21, 30, 91, 131, 257
Ancient texts 34, 45, 59, 61, 93, 223, 257
Ancient world 3–6, 20–26, 97,

101–110, 117, 121, 123, 126–136, 140, 144, 150–151, 174, 177, 185–188, 200, 203, 211, 229, 257–266
Anti-Christian 173
Anti-Semitism 118–119, 148–149, 162–163, 212
Antichrist 61
Apes 1, 232
Archives 88
Aristotle 124
Ark of the Covenant 60, 67, 82, 261
Armageddon 60–61, 67, 80–81
Artificial selection 149
Astronomy 25, 48, 53, 107, 112, 125
Atheist 4, 29, 34, 38, 46–48, 53, 56, 101, 122–127, 188, 242, 249
Atheistic 5–9, 15, 19–22, 44–58, 89, 99, 122–137, 142–148, 157, 159, 164–167, 171, 173, 182, 185, 200, 226, 231, 250–252, 258
Atheistic zealots 258
Axis of the Earth 38, 43, 50, 145, 146
Axis Powers 145, 148

B

Ba'-al 203, 204
Baalbek 33–34
Babylon 60, 75, 213
Babylonian 73, 88, 113, 116
Beirut 179, 192, 206
Belfast 105, 192, 206, 258
Berlin 126, 180
Bible 14, 19–23, 89, 93, 101–102, 122, 159, 168, 171, 181, 191, 195, 197, 211, 240–241, 245–247, 258, 261
Birth control 174–176, 230
Born Again 209–210, 253, 263–264
Bosnia 105, 192, 206
Britain (see also England) 139, 145, 148, 161, 163
Buddha 246

C

Caesarea 33–34
Capitalism 135, 143, 160, 165, 172, 180
Catholic 114, 117, 127, 161, 166, 172, 176, 191, 199, 206, 233–238, 240, 242, 246, 257, 262
Chiang Kai-Shek 147
Chichen Itza 93, 97
China 147, 165, 177
Chosen People 115, 117, 160, 162, 194–198, 214
Christ-killers 116, 118
Christian 3, 10, 19, 23, 191, 194, 197, 223, 233, 239, 244, 253, 263
Christian leaders 68, 205, 233, 261
Christian Right 175, 184, 192, 240
Christianity 4, 21, 61, 67, 137, 140, 181, 203, 205, 257, 261
Christians 66, 114, 116, 118, 140, 145, 157, 162, 166, 168, 183
Church 3–9, 21, 23, 98, 113, 116, 118, 123–124, 126, 129–131, 136, 161, 174, 191, 226, 238, 240, 242, 257
Church scientists 3, 124, 130
Civil war 69–79, 82, 86–89, 108–110, 133, 176, 191, 259–260
Clergy 4, 7, 16, 23–24, 26, 46, 53, 126, 136, 172, 185, 205, 214, 218, 229, 237–239, 241, 243, 247, 263
Clinton, Bill 180, 230, 250–252
Cloning 116, 158, 231
Cold War 23, 160, 165, 176, 178, 180, 184, 191, 239
Columbus 26, 59, 121
Communism 105, 132, 136, 142, 145, 155, 159, 164, 171, 177, 180–182, 258
Confucius 246
Congress of Vienna 134, 138
Copernicus, Nicholas 8, 124
Creator 6, 59, 181, 243, 262
Crusades 117, 121, 124, 132, 160, 185
Cult 4–7, 14, 18–24, 30, 48, 89, 99, 101, 123, 126, 129, 130–137, 157, 159, 160, 167, 182, 191, 199, 200, 213, 222, 226–227, 231, 246, 249, 252–253, 262, 264
Czar Nicholas II 106, 139, 140, 142, 178

D

Daily sacrifice 74, 260
Daniken, Erich von 49, 53, 55–56, 102, 132, 185, 227–228
Dark ages 3, 20, 22, 26, 51, 112, 123, 126, 128, 177, 188, 253
Darwin, Charles 4–5, 9–12, 14, 16–17, 21–23, 30, 40, 46, 52, 55, 130, 149–159, 167, 181–182, 185, 213, 223–224, 226, 259

Darwinian 3, 4, 10, 14, 52, 56, 100, 137, 148, 150, 154–156, 158–159, 169, 182, 199, 212–213, 223–224, 228, 258

Darwinian Holocaust (see also Holocaust) 132, 148, 152, 160, 231, 258

Davidson, David 41, 43–44, 50, 56

Death 2–3, 8, 103, 232, 240, 246, 249, 263

Demigod 56–57, 60, 91, 105–107, 146, 185

Denton, Michael 9–16, 56, 132, 159, 181, 185, 208, 227–228

Descent Of Man 15, 150, 152–153, 157–158

Diaspora 113

Disciples 62–69, 91, 102, 128, 197–199, 205, 230

Druids 30, 114

E

Eden 1, 59, 90, 124, 258

Egypt 33, 35, 38, 45, 48, 51, 57, 60, 86, 90, 195–196

Egyptian 22, 25, 114

Egyptians 35, 38, 44, 49–52, 55, 58

Elijah 203–206

Emperor Constantine 116

Emperor Hirohito 213

Emperor Nero 85–86, 252

Emperor Titus (see Titus)

Emperor Vespasian 108–111, 118

Empiricist 126–128, 130–131, 259

Engels, Friedrich 136

England (see also Britain) 106, 119, 121, 133–134, 149, 266

Equinox 25, 30, 43, 49, 50, 93

Evolution 1–18, 21, 23, 30, 46, 52, 127–132, 149, 154, 158, 167–168, 171–172, 181–182, 200, 208, 214,

241, 246, 258–259, 264

Extraterrestrial 1, 3–4, 22, 53, 55, 91

F

False Apostles 114, 203

False Messiah 260

Famine 58, 69, 71, 80, 89, 123

Fanaticism 127–128, 159, 185, 213–214, 232

Fascism 136, 143, 159–160, 177, 258

Fossil 11, 30, 130

France 106, 121, 133–134, 138, 144, 160–161

Fundamentalist 10, 158, 168, 172, 179, 248–249

G

Galileo 3–4, 25, 56, 124–125, 184, 222, 231

Genetic engineering 231

Gentiles 197–199

Geocentric 7, 12, 14, 124–125, 159

Germany 121, 134, 138–139, 144, 146, 150, 153, 157, 176

Giza 35, 40

God of Abraham 57–61, 73, 82, 103, 117, 197, 203–204, 224, 260

God-men 106

Gods 1, 5–6, 20–29, 34, 39, 46, 51–58, 61, 65, 78, 81, 89, 91, 97, 102–112, 117, 121, 129, 132, 135, 177, 187, 203, 259, 266

Golden Calf 195

Goliath 57

Goths 144, 163

Graham, Rev. Billy 16, 218, 256

Great Pyramid 26, 35, 38, 45, 47–48, 50, 61, 122, 251

Great Year 43

Gypsies 153, 156–157, 213

H

Heathens 67, 114, 196, 199
Hegel, Georg Wilhelm Friedrich 127, 131, 259
Heliocentric 8, 12, 124
Hell 164, 175, 232, 242, 246
Hieroglyphics 22, 50, 53, 107
High Priest 76, 83, 109
Hitler, Adolph 61, 100, 132, 134, 144–149, 153–163, 213–214
Hollywood 168, 173, 229, 239, 243, 246
Holocaust (see also Darwinian Holocaust) 17, 80, 160, 172–178, 185, 191, 209–214, 231, 239, 259
Holy House (see Temple)
Homer 19, 56–58, 89, 122, 241
Human sacrifice 225–226
Humanism (see Secular Humanism)
Humanists 168, 175, 244, 249

I

Iliad 19, 122
Inca 1, 56
Inquisition 121, 124, 149, 157
Iron Curtain 161
Islam 117, 130, 143, 179, 182, 191, 193, 200, 205, 255, 257
Israel 1, 22, 57, 59, 60, 68, 82, 88, 91, 101, 105, 114–115, 162–164, 174, 178, 187, 191–197, 203–214, 235, 247, 261
Israelites 51–68, 114, 194–198, 204, 211, 234, 259–261
Italy 106, 134, 138–139, 143–146, 213, 235, 237–238

J

Japan 106, 108, 139, 145–146, 148, 213, 237, 242

Jehovah's Witnesses 45, 154, 193
Jeremiah 113, 196, 198
Jerusalem 26, 39, 60–73, 81–91, 102–117, 120, 159–160, 164, 188, 194, 198, 200, 203, 206, 239, 260–261
Jesus 57, 61–71, 76–77, 83, 89, 90–92, 101, 107, 114–120, 162, 164, 172, 193–194, 197–198, 200, 205–206, 213, 219, 233, 247, 260, 264
Jews 17, 19, 63–69, 71–79, 83–89, 109–120, 140, 145, 148, 153–164, 178–188, 191–199, 203, 212, 214, 239–241, 260
Jihad 6–7, 105, 137, 178, 187, 222, 227, 244, 250
John, The Apostle 92, 194
Josephus 33, 56, 59, 62–68, 70–79, 81–88, 91, 105, 109, 258
Judah 196–199
Judaism 82, 114, 117–118, 130, 143, 160, 162, 191, 199, 203, 205, 233, 257, 260
Judea 108
Judeo-Christian 81, 113, 173, 206–207, 219–227, 231, 239
Judeo-Christianity 2, 89, 131, 212, 232, 234, 243–244, 246
Judgment 95, 225, 227

K

Kepler, Johannes 125
King David 80, 92, 103, 186, 188, 267
King Henry VIII 121
King Herod 33, 61, 69
King Louis XVIII 134
Kingdom 69, 195, 197–199
Kushner, Rabbi Harold S. 207–210, 221, 224, 227

L

Lamarck, Jean Baptiste 130
Leo, John 224, 227–228, 231, 252
Liberal 134, 164, 166, 174, 177, 180, 192, 239–240, 247
Lincoln, Abraham 191, 260
Luther, Martin 121, 127

M

Magna Carta 133
Mao Tse-tung 132, 147
Marx, Karl 128, 136, 140–142, 167, 181
Maya 1
Mecca 255
Meek 60, 91–92, 101, 107, 186
Mengele, Josef 158
Menzies, Robert 40, 44, 50
Messiah 45, 62–68, 83–92, 101–118, 140, 159, 162, 182, 193–194, 198, 200, 203, 212, 260
Mohammed 117, 193, 199
Moses 56, 60, 118, 195, 219, 247
Muslims 162–163, 179, 191–193, 197, 199, 255
Mussolini, Benito 100, 134, 143–144, 146, 159, 163
Myth 1–22, 34, 45, 93, 122, 126, 129–132, 158–159, 170–171, 199, 214, 218–222, 231, 238, 241, 246–247, 258, 266–267
Mythology 32, 56, 101

N

Napoleon 100, 133–134, 138, 143–144
Nationalism 134, 138, 143
Natural Law 2, 43, 127, 129
Nazi 16, 149–158, 162, 166, 176, 213, 240
Neanderthal 24, 240

Nebuchadnezzar 113
New Covenant 196
New Testament 19, 62–63, 66, 70, 81, 90, 92, 103, 194, 196, 198, 205, 260–261
New World 112, 122, 177
Noah 39
Noble Lie 218–221, 223, 226

O

Odyssey 19, 122
Old Testament 63, 69, 83, 92, 103, 193, 196, 203, 260

P

Pagan gods 57
Paleontology 9, 13, 131
Palestine 115, 163
Papal 5, 106, 117, 238
Paradigm 6–17, 22, 26, 52–53, 92, 100–102, 132, 186, 207–208, 218, 221–228, 231, 240, 250, 253, 258
Paradigm lock 93, 100, 217
Paris 126–128, 134
Patrician 109
Paul, The Apostle 24, 56, 65–66, 91, 103, 198–199
Pearl Harbor 148
Pehme, Kalev 158
Pelota Ballcourt 95
Pestilence 69, 71, 89
Peter, The Apostle 81, 198–199
Pharaoh 45, 57
Pisar, Samuel 212
Planned Parenthood 230, 240, 244
Pope 117, 119, 135–136, 138–139, 142, 172, 237
Prayer 174, 205–209, 212, 229, 233, 246, 264
Precession 43
Primogeniture 106–107, 266

Ptolemy 12, 124
Punctuated equilibrium 11–13, 30
Pyramid 35, 38, 40–41, 43–44, 47–48, 93–95, 97

R
Rabbi Akiba 115
Rapture 65–66, 102, 203
Reagan, Ronald 180, 233
Religion 2–9, 15–20, 27–35, 89, 101–118, 124, 128–131, 135, 137, 157, 174–188, 200, 206, 208, 211–212, 218, 222, 226, 228, 230, 232, 235, 239, 245, 249, 256–258, 263
Renaissance 4, 122, 124, 129
Resurrection 219, 247
Revolution 15, 112, 122, 126, 128, 133, 137, 139–140, 142–147, 160, 171–181, 186, 259, 267
Roe vs Wade 174
Romans 51, 56, 70–78, 82–88, 101, 113–120, 163, 194
Rome 90, 108, 114–123, 139, 143, 154, 229, 235, 238, 252, 258
Rue, Prof. Loyal D. 218, 221, 223–224, 227
Russia 106, 133, 139–148, 155, 160, 164, 178, 180, 182

S
Sanctuary 71–72, 75, 78–79, 82, 87
Satan 59, 81, 89–99, 101–103, 121, 136, 186–188, 233
Satanic 5, 47, 59, 89, 101, 116–121, 136, 169, 183–187, 257, 262
Schliemann, Heinrich 19–20, 24–26, 53, 56, 58, 101
Scopes Trial 168, 170, 251
Second coming 45, 63, 65, 68, 89, 102, 106, 193, 200
Second life 263

Secular Humanism 123, 129–130, 137, 143, 172, 174, 192, 226, 229, 244, 251
Seditious 73, 77–78, 80, 83–84, 89, 109, 121
Serpent 57, 93, 95, 97
Sexual Revolution 174
Skulls 97
Slavery 21, 60, 73, 76, 135, 185, 225, 259
Slaves 34, 88, 91, 106–107, 185
Smyth, Prof. Charles Piazzi 38–40, 44, 49, 56
Solar system 7–8, 12, 26, 41, 48, 59, 93, 107, 124, 159, 262
Solstice 116
Spain 119, 121, 124, 138, 149
Sphinx 54
Stalin, Josef 159, 161
Star 63
Stonehenge 30, 32, 34
Sun Yat-Sen 147
Suspended prophecy theory 67, 82

T
Tabernacle 113
Tacitus 64–66, 81, 91
Talmud 113
Taylor, John 38–41, 44, 49–50, 56
Temple 39, 60–69, 71–87, 115, 198, 260
Temples 23, 30, 55, 106–107, 187
The Castle 93, 96
Titus 73–74, 78–79, 82–84, 87, 108
Tompkins, Peter 35–48, 50–57, 102, 132, 185, 227–228
Troy 19, 26, 34, 53, 58, 89

U
Uniformitarian Principles 130
United Nations 162–163, 213–214

V

Vatican 117, 121, 138–139, 142, 145, 161, 171, 182
Virgil 56–58, 122, 241
Voltaire 126

W

World War I 138, 140, 144–146, 163
World War II 148, 160–161, 165–166, 178, 183

Y

Yom Kippur 233

Z

Zealots 22, 30, 71–72, 76, 167, 183
Zionist 162–163